An Introduction to Data Analysis using Aggregation Functions in R

Simon James

An Introduction to Data Analysis using Aggregation Functions in R

 Springer

Simon James
School of Information Technology
Deakin University
Burwood, VIC, Australia

ISBN 978-3-319-83579-2   ISBN 978-3-319-46762-7   (eBook)
DOI 10.1007/978-3-319-46762-7

This Springer imprint is published by Springer Nature
The registered company is Springer International Publishing AG
The registered company address is: Gewerbestrasse 11, 6330 Cham, Switzerland

Preface

This introduction to the area of aggregation and data analysis is intended for computer science or business students who are interested in tools for summarizing and interpreting data, but who do not necessarily have a formal mathematics background. It was motivated by our need for such a text in postgraduate data analytics subjects at my university; however it could also be of interest to undergraduate data science students. By reading through the chapters and completing the questions, you will be introduced to some of the issues that can arise when we try to make sense of data. It is hoped that you will gain an appreciation for the usefulness of the results established in the field of data aggregation and their wide applicability. I personally find the study of aggregation functions to be very enjoyable and fulfilling, as it allows for a nice mix of theoretically interesting results and practical applications. The rise of data analytics and the growing need for companies to learn more from their data also make the need for intelligent data aggregation techniques more indispensable than ever.

The overall aim of this book is to allow future data analysts to become aware of aggregation functions theory and methods in an accessible way, focusing on a fundamental understanding of the data and summarization tools that complements any study in statistical or machine learning techniques. To this end, included in each of the chapters is an R tutorial giving an introduction to commands and techniques relevant to the topics studied. These tutorials assume no programming background and will give you some exposure to one of the most widely adopted (and freely available) computing languages used in data analysis.

For a more in-depth overview of aggregation functions, there are some great existing monographs, including the following to name a few.

- Beliakov, G., Bustince, H. and Calvo, T.: A Practical Guide to Averaging Functions. Springer, Berlin, New York (2015)
- Beliakov, G., Pradera, A. and Calvo, T.: Aggregation Functions: A Guide for Practitioners. Springer, Heidelberg (2007)
- Gagolewski, M.: Data Fusion. Theory, Methods and Applications. Institute of Computer Science, Polish Academy of Sciences (2015)

- Grabisch, M., Marichal, J.-L., Mesiar, R. and Pap, E.: Aggregation Functions. Cambridge University press, Cambridge, (2009)
- Torra, V. and Narukawa, Y.: Modeling Decisions. Information Fusion and Aggregation Operators. Springer, Berlin, Heidelberg (2007)

Whilst focusing on established results, I hope that the following pages will offer a fresh perspective on the recent trends in aggregation research. In this respect I am indebted to a number of colleagues and researchers who lead this field, many of whom I have been fortunate enough to meet or collaborate with over the span of my short career. In particular, I would like to acknowledge Humberto Bustince, Radko Mesiar, and Michal Baczynski, who have all made me feel very welcome when visiting their institutions and participating in conferences overseas. Special mention should be made regarding the works of Michel Grabisch and Ronald R. Yager on fuzzy integrals and OWA operators respectively, which have been a huge influence in my understanding of these areas (and so I hope I do them justice in the short introductions provided). I am grateful to my friend and ecology research expert, Dale Nimmo, who has helped identify a number of interesting applications aggregation theory. A huge thank-you to Marek Gagolewski for our recent discussions and for his extensive comments on drafts of this text; I am looking forward to our future research endeavors! Finally (on the academic side), I would like to thank Gleb Beliakov for continuing to be a mentor to my research. Certainly the majority of these chapters have been influenced by what you have taught me and our work together.

Lastly I would like to acknowledge the support of family (in particular my Mum, who proofread some chapters of this book), friends (especially Rachel, who has been a great support while I've been working on this book in addition to her usual honest and constructive comments), and my work colleagues (a very busy 2016 was made much more bearable by having people like Lauren, Elicia, Guy, Tim, Lei, Crystal, Menuri, Gang, Shui, Vicky, Sutharshan, and Michelle to talk to).

Melbourne, VIC, Australia   Simon James
August 2016

Contents

1 Aggregating Data with Averaging Functions 1
  1.1   The Problem in Data: Average Returns 2
  1.2   Background Concepts 4
       1.2.1   What is a Function? 4
       1.2.2   Multivariate Functions and Vectors 7
  1.3   The Arithmetic Mean 10
       1.3.1   Definition 10
       1.3.2   Properties 12
  1.4   The Median 14
  1.5   The Geometric and Harmonic Means 16
       1.5.1   The Geometric Mean 16
       1.5.2   More Properties 18
       1.5.3   The Harmonic Mean 19
  1.6   Averaging Aggregation Functions 21
  1.7   Uses of Aggregation Functions 23
       1.7.1   Sporting Statistics 24
       1.7.2   Simple Forecasting 24
       1.7.3   Indices of Diversity and Equity 25
  1.8   Summary of Formulas 25
  1.9   Practice Questions 26
  1.10 R Tutorial 27
       1.10.1 Entering Commands in the Console 27
       1.10.2 Basic Mathematical Operations 27
       1.10.3 Assignment of Variables 28
       1.10.4 Assigning a Vector to a Variable 29
       1.10.5 Basic Operations on Vectors 29
       1.10.6 Existing Functions 30
       1.10.7 Creating New Functions 32
  1.11 Practice Questions Using R 33
  References 34

2 Transforming Data 37
  2.1   The Problem in Data: Multicriteria Evaluation 38
         2.1.1   Which is Better: Higher or Lower? 39
         2.1.2   Consistent Scales 39
         2.1.3   Differences in Distribution 39
  2.2   Background Concepts 41
         2.2.1   Arrays and Matrices ($\mathbf{X}$) 41
         2.2.2   Matrix/Array Entries (x_{ij}) 41
         2.2.3   Matrix/Array Rows and Columns ($\mathbf{x}_i, \mathbf{x}_j$) 42
  2.3   Negations and Utility Transformations 42
  2.4   Scaling, Standardization and Normalization 46
  2.5   Log and Polynomial Transformations 50
  2.6   Piecewise-Linear Transformations 52
  2.7   Functions Built from Transformations 56
  2.8   Power Means 57
  2.9   Quasi-Arithmetic Means 60
  2.10   Summary of Formulas 61
  2.11   Practice Questions 62
  2.12   R Tutorial 63
         2.12.1   Replacing Values 63
         2.12.2   Arrays and Matrices 64
         2.12.3   Reading a Table 65
         2.12.4   Transforming Variables 67
         2.12.5   Rank-Based Scores 67
         2.12.6   Using if() for Cases 68
         2.12.7   Plotting in Two Variables 69
         2.12.8   Defining Power Means 71
  2.13   Practice Questions Using R 72
  References 72

3 Weighted Averaging 75
  3.1   The Problem in Data: Group Decision Making 76
  3.2   Background Concepts 78
         3.2.1   Regression Parameters 78
  3.3   Weighting Vectors 79
         3.3.1   Interpreting Weights 81
         3.3.2   Example: Welfare Functions 81
  3.4   Weighted Power Means 82
  3.5   Weighted Medians 84
  3.6   Examples of Other Weighting Conventions 85
         3.6.1   Simpson's Dominance Index 85
         3.6.2   Entropy 86
  3.7   The Borda Count 87
  3.8   Summary of Formulas 88
  3.9   Practice Questions 89

3.10 R Tutorial 91
    3.10.1 Weighted Arithmetic Means 91
    3.10.2 Weighted Power Means 91
    3.10.3 Default Values for Functions 92
    3.10.4 Weighted Median 92
    3.10.5 Borda Counts 93
3.11 Practice Questions Using R 95
References 95

4 Averaging with Interaction 97
4.1 The Problem in Data: Supplementary Analytics 98
4.2 Background Concepts 99
    4.2.1 Trimmed and Winsorized Mean 99
4.3 Ordered Weighted Averaging 101
    4.3.1 Definition 101
    4.3.2 Properties 102
    4.3.3 Special Cases 102
    4.3.4 Orness 103
    4.3.5 Defining Weighting Vectors 104
4.4 The Choquet Integral 108
    4.4.1 Fuzzy Measures 108
    4.4.2 Definition 110
    4.4.3 Special Cases 111
    4.4.4 Calculation 111
    4.4.5 Examples 113
    4.4.6 Properties 115
    4.4.7 The Shapley Value 116
4.5 Summary of Formulas 117
4.6 Practice Questions 118
4.7 R Tutorial 120
    4.7.1 OWA Operator 121
    4.7.2 Choquet Integral 122
    4.7.3 Orness 123
    4.7.4 Shapley Values 125
4.8 Practice Questions Using R 125
References 126

5 Fitting Aggregation Functions to Empirical Data 129
5.1 The Problem in Data: Recommender Systems 130
5.2 Background Concepts 134
    5.2.1 Optimization and Linear Constraints 135
    5.2.2 Evaluating and Interpreting a Model's Accuracy 139
    5.2.3 Flexibility and Overfitting 142
5.3 Learning Weights for Aggregation Functions 143
    5.3.1 Fitting a Weighted Power (or Quasi-Arithmetic) Mean 143
    5.3.2 Fitting an OWA Function 144
    5.3.3 Fitting the Choquet Integral 145

5.4   Using Aggregation Models for Analysis and Prediction   146
       5.4.1   Comparing Different Averaging Functions   149
       5.4.2   Making Inferences About the Importance of
               Each Variable   155
       5.4.3   Make Inferences About Tendency Toward
               Lower or Higher Inputs   156
       5.4.4   Predicting the Outputs for Unknown/New Data   156
5.5   Reliability   157
       5.5.1   Do Our y_i Values Represent the 'Ground Truth'?   157
       5.5.2   Are We Evaluating an Approach or the Model Itself?   158
5.6   Conclusions   159
5.7   Summary of Formulas   160
5.8   Practice Questions Using R   161
References   162

6   Solutions   163
    Aggregating Data with Averaging Functions: Solutions   163
    1.9   Practice Questions   163
    1.10   Practice Questions Using R   165
    Transforming Data: Solutions   168
    2.11   Practice Questions   168
    2.13   Practice Questions Using R   172
    Weighted Aggregation: Solutions   176
    3.9   Practice Questions   176
    3.11   Practice Questions Using R   180
    Averaging with Interaction: Solutions   182
    4.6   Practice Questions   182
    4.7   Practice Questions Using R   190
    Fitting Aggregation Functions to Empirical Data: Solutions   192
    5.6   Practice Questions Using R   192

Index   197

Chapter 1
Aggregating Data with Averaging Functions

Data can be overwhelming. We might be able to mentally make sense of a set of 10 or so numbers, but beyond this we will usually benefit from some assistance from technology. Of course, these days the sheer volume of data we are able to collect is sometimes so high that even the fastest computers and modern software can't process and analyse it within a useful timeframe. Understanding data hence usually involves summarizing it, however since we inevitably lose information in the process, we want to make the summary as useful as possible.

In making decisions based on available data when there is some degree of uncertainty about how our choice will unfold, we often rely on past experience and a sense of what is 'normal' or 'expected' with respect to each of the potential outcomes. This can be from our own personal experience, for example, knowing how long it takes to drive to work depending on the time we leave, however sometimes we will be reliant (at least to some extent) on second-hand data. Talent scouts for the NBA can't attend every college basketball game (there are some 350 odd teams in Division 1 alone) and so statistics like average points per game, minutes played, winning percentage, frequency of injury, and so on, will undoubtedly play a role in getting a feel for a player's expected performance. Furthermore, it is usually preferable to base our decision on numerous aggregated values because a single datum by itself will never be informative.

The most commonly employed tool for averaging data is the **arithmetic mean**, which is the sum of all numbers divided by how many there are, usually referred to as the 'mean' or the 'average'. In this chapter we will see that this function is just one of many potential averages that can be used to summarize data, and further, that in some situations it is neither useful nor appropriate to use the arithmetic mean.

© Springer International Publishing AG 2016
S. James, *An Introduction to Data Analysis using Aggregation Functions in R*,
DOI 10.1007/978-3-319-46762-7_1

Chapter Objectives

- To introduce the concept of averaging aggregation functions along with the arithmetic, harmonic and geometric means
- To develop intuition about when it is appropriate and/or useful to use each of these means
- To introduce basic commands and syntax in R

1.1   The Problem in Data: Average Returns

When it comes to evaluating the attractiveness of an investment, we might know nothing except its yearly returns. Below are the yearly returns for two potential investments, *Taida* and *Littlebig*.

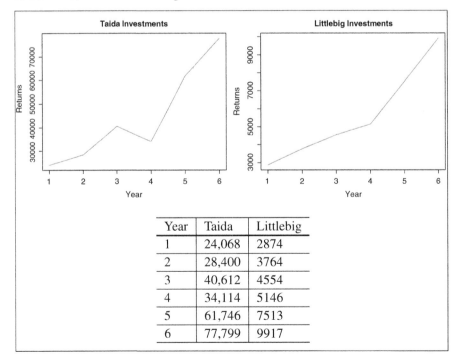

Year	Taida	Littlebig
1	24,068	2874
2	28,400	3764
3	40,612	4554
4	34,114	5146
5	61,746	7513
6	77,799	9917

Obviously there will be a lot of variability and chance with regard to future returns, but let's say we want to choose the more sound option based on the performance over the last 5 years.

In order to compare companies of different sizes, we are more interested in the *rate* of return per dollar invested, and rather than trying to comprehend a trend over 5–10 years it becomes a more manageable task if we summarize this data. However something counter-intuitive happens when we average these rates of return. Let's first calculate the rate of return (percentage increase on the previous year) for each of these companies.

Year	Taida	(Rate %)	Littlebig	(Rate %)
1	24,068	–	2874	–
2	28,400	18	3764	31
3	40,612	43	4554	21
4	34,114	−16	5146	13
5	61,746	81	7513	46
6	77,799	26	9917	32
Average		30.4		28.6

Adding the rates and dividing by 5 gives us 30.4 % and 28.6 % respectively, so it seems like *Taida* performs better 'on average'. But how do we interpret these averages?

If we are told that the average cost of a wedding is $65,000, then we know this is based on the total cost of all the weddings added together divided by the number of weddings. If every couple spent the same, then they would each spend $65,000 (of course we know that some will spend less and some will spend much much more!). In hearing this, we also might form an idea that this is how much a 'normal' wedding costs, and so if we are planning to get married then this is how much we should expect to pay.

The impression that the 30.4 and 28.6 summaries might give us is that this rate of return is what we might expect each year. However if this were the case, our returns would look like this:

Year	Taida	Expected 30.4 % pa	Littlebig	Expected 28.6 % pa
1	24,068	24,068	2874	2874
2	28,400	31,385	3764	3696
3	40,612	40,926	4554	4753
4	34,114	53,368	5146	6112
5	61,746	69,592	7513	7860
6	77,799	90,748	9917	10,108
Difference		12,949		191

In both cases, the final amount that would have been expected is higher than the actual final investment worth. In the case of *Taida*, the difference is substantial, and so we might question whether summarizing the rates of return in this way is misleading and indeed whether it is useful as a basis for making our decision.

The main issue leading to this difference is that the rates are related to the data through *multiplication*, rather than through *addition*. In particular, we can link the starting amount with *Taida* to the final amount via the rates with the following equation.

$$30{,}500 \times 1.18 \times 1.43 \times 0.84 \times 1.81 \times 1.26 = 77{,}799.$$

If the rates of return were the *same across each year*, the multiplier should be approximately 1.264, which means a rate of 26.4 %. This is not a world away from our 30.4 % figure, however it is different enough that our final investment is overestimated by almost 17 %. In the case of *Littlebig*, we only get about a 2 % difference.

As we will see, the overestimation due to using the simple average in this scenario will become more pronounced the more the rates vary. Although this method of summarizing the rates is flawed at its basis, it is none-the-less still used widely for this purpose. Of course, if *Taida* or *Littlebig* want to promote their success, it sounds much better to quote the average values here.

In fact, the value of 1.264 or 26.4 % can also be referred to as an **average** or a **mean**, and is obtained using a function called the **geometric mean**. We will be focusing on this and a few other functions as alternatives to the arithmetic mean (what is usually referred to as the 'mean' or 'average') and look at how they are calculated and interpreted. We will first need to familiarize ourselves with some terminology and refresh some mathematical concepts.

1.2 Background Concepts

Although you will undoubtedly have come across the concepts of averaging before, here we will need to clarify some notation and terminology so that we can start to think about and describe the arithmetic mean (and other functions) with respect to formal properties and characteristics.

1.2.1 What is a Function?

The formal definition of a function has arisen from various developments in different areas of mathematics [14]. The one we might be most familiar with is given in terms of a value x, where the function, denoted $f(x)$ or y, gives the resulting value for every instance of x.

An example is

$$f(x) = 4x + 2.$$

Most of the time, the x is assumed to be a *real* number (i.e. every value that we could represent on a number line, like $10, -5, 3.75, \frac{3}{11}$ and even numbers like $\pi \approx 3.1416$ and $\sqrt{2} \approx 1.4142$). So we have,

$$
\begin{aligned}
f(2) &= 4 \times 2 + 2 &=& \quad 10 & \text{(input 2, output 10)} \\
f(-3) &= 4 \times -3 + 2 &=& \quad -10 & \text{(input} -3, \text{output} -10) \\
f(\pi) &= 4 \times \pi + 2 &=& \quad 4\pi + 2 & \text{(input } \pi, \text{output} \approx 14.57) \\
f(-\sqrt{7}) &= 4 \times -\sqrt{7} + 2 &=& -4\sqrt{7} + 2 & \text{(input} -\sqrt{7}, \text{output} \approx -8.58)
\end{aligned}
$$

However, in general x could be something else and the expression of the function could take a complex number (i.e. involving the square root of a negative number), a vector of inputs like $x = \langle 3, 2, 0, 8 \rangle$ or a qualitative descriptor like "large". If this were the case, we would need to make sure we know how to define $4x$ (the number four, multiplied by our input) and we would need to know how to add 2.

A clearly stated expression like $4x + 2$ can actually be very uncommon. Consider going to the airport and parking. The price will often be something like the following:

Hours	Price [$]
0–0.5	$4.00
0.5–1	$7.00
1–3	$10.00
3–5	$18.00
Per day	$25.00

The price here is a 'function' of the time we park for. We could write it formally as a mathematical expression, however airport customers might have trouble getting a clear idea of the costs in this form. Expressed as a piecewise function, the parking costs based on hours of stay $Cost(h)$ would be,

$$
Cost(h) = \begin{cases}
4, & 0 \le h < 0.5, \\
7, & 0.5 \le h < 1, \\
10, & 1 \le h < 3, \\
18, & 3 \le h < 5, \\
25 \left\lceil \frac{h}{24} \right\rceil, & \text{otherwise.}
\end{cases}
$$

Notation Note Ceiling $\lceil \cdot \rceil$

The special brackets $\lceil \cdot \rceil$ denotes the 'ceiling' function. It means we go upwards to the next integer (unless we have an integer already, in which case we keep that value). If h were 36, then we get $36/24 = 1.5$ and the next integer up is 2 so our total cost would be $2 \times 25 = \$50$. However if h were 72, then we get $72/24 = 3$ and so we would have $3 \times 25 = \$75$.

There is no extra information in this expression than what is in the table, i.e. the table is as valid an expression of our cost function as our equation that makes use of mathematical notation. The key idea is that for any *input* the function deterministically produces a single *output*.[1] It would be very confusing if our parking table looked like this:

Hours	Price
0–0.5	Sometimes \$4, sometimes \$8
0.5–1	Between \$7.00 and \$20.00
1–3	Either \$10 or \$12
3–5	\$18.00
Per day	\$25.00

Of course, the price may change tomorrow, but for when we enter the carpark, we want the output value for any input x to be unambiguous.

Example 1.1. Does the following rule-based system (for controlling the orientation of a ground-based drone when obstacles are close) define a function?

Obstacle position (x) (°)	Set drone direction
0–90	$x - 100$
90–270	0
250–300	10
300–360	$x + 100$

Solution. No. There are two issues that mean we do not have a function here. The first is that there is an overlap in the intervals so that between 250° and 270°, we don't know whether the drone should have its direction set to 0 or 10.

The second, which is only minor and more of a technicality, is that the drone does not know what to do on the boundary of each interval, i.e. at 90°, the first interval says it should set its direction for -10 while the second interval says zero. If we changed the third interval to 270–300 and then clarified that on

---

[1]This broader definition of a function extends to computer functions and algorithms too. An object, file, image, etc., is submitted and an output is produced.

the boundary we should go with the lower bin, then we would have a function.
Note too that 0 is the same as 360 so this case would need to be determined too.

1.2.2 Multivariate Functions and Vectors

A multivariate function extends the idea of functions of a single variable like $f(x)$ to allow for multiple arguments. So instead of an input like '3', we have a set of inputs, e.g. $3, 4, 2, 1.5$ all at once. We sometimes refer to this set of inputs as a *vector*. Vectors are often introduced as measurements that have both direction and magnitude, for example, windspeed is usually given in terms of its strength and direction (e.g. 10 km/h south-westerly). 'Direction' is easy enough to understand when we're talking about two and three dimensions, however we can also have vectors with 5, 10, 100 or billions of arguments. We will use the term 'vector' to denote any set of arguments where the position of each entry usually carries with it some kind of meaningful interpretation and we will use angular brackets $\langle\rangle$. Some examples include:

- **position vectors** (of length 2 or 3), e.g. $\langle 3, 4, 2 \rangle$ denotes the co-ordinates of a point where the value in the x dimension is 3, in the y dimension is 4, and the z dimension is 2 (of course, the labels x, y, z are arbitrary);
- **sets of ratings**, e.g. a movie on iMDb might have received scores from 7 people, which can be represented as the vector $\langle 6, 7, 8, 8, 8, 3, 7 \rangle$. In this case, each input represents the rating given to the movie by one person;
- **measurements for varying attributes**, e.g. in medical research we might have a record which describes each subject in terms of their height, weight, heart rate, number of hospital visits, and presence/absence of a disease. For example, one such description is $\langle 172, 67, 81, 3, 0 \rangle$. In this case, each dimension is measured on a different scale;
- **historical data**, e.g. in forecasting the day's sales at a fast-food store, we might be interested in the sales from the same weekday over the last 3 weeks. Each of the entries in the vector $\langle 3456, 2349, 3511 \rangle$ then tells us the total sales on those three separate days.

We will denote generic vectors using bold letters. Typically we will have $\mathbf{x} = \langle x_1, x_2, \ldots, x_n \rangle$ where '$\mathbf{x}$' refers to the whole vector and x_1 refers to the first component, x_2 to the second and so on. The length of the vector (usually denoted by n) is the total number of elements. In some applications, this will not vary (e.g. for position vectors in three dimensions we will always have $n = 3$), whereas in some cases we might need to deal with vectors of different length, e.g. with iMDb ratings some movies will have more ratings than others.

For a multivariate function, we can use the notation $f(\mathbf{x})$ instead of $f(x)$ and at times we may distinguish between different functions. For example, we will use

AM($\mathbf{x}$) to refer to the arithmetic mean and we will usually write $f(x_1, x_2, x_3)$ rather than $f(\langle x_1, x_2, x_3 \rangle)$.

Notation Note $f(x_1, x_2, x_3)$ rather than $f(\langle x_1, x_2, x_3 \rangle)$

Technically interpretations arising from these representations should differ in that $f(x_1, x_2, x_3)$ indicates a multivariate function that takes 3 real numbers as input and $f(\langle x_1, x_2, x_3 \rangle)$ is a function that takes a single input, however the input is 3-dimensional. For our considerations there is usually no practical difference so we will opt for the former (since it is less cumbersome), however for programming functions in R, it will become a point of difference (see Sect. 1.10.7). It is usually easier to write functions using `F <- function(x) {...}` (where x can be a vector of any length), rather than `F <- function(x,y,z) {...}` if we are intending x, y, z to be real arguments.

Vectors can also be thought of as either the rows or columns of a table. When thinking ahead to how we will use and interpret our data, it is sometimes helpful to think of how we are going to store it, organize it, and how we might interpret the resulting cells of a table or spreadsheet.

In all of our chapters we will be concerned with multivariate functions, which take a vector as input and result in a single output. We will be dealing primarily with vectors of real numbers as inputs and outputs also given as real numbers. The individual arguments of the input vectors and the outputs will also usually have the same range.

Notation Note Multivariate functions

Formally we can indicate preliminary information about our function as follows:

$$f : [0, 1]^3 \rightarrow [0, 1],$$

which means that f is a function of 3 arguments, which all have values in the interval $[0, 1]$, and whose output is in the interval $[0, 1]$. The colon ':' can be read "such that" and the arrow '$\rightarrow$' can be read as "maps to". Instead of $[0, 1]^3$ we could write $[0, 1] \times [0, 1] \times [0, 1]$ (this notation indicates the "Cartesian product" of sets), however we could also have arguments that range over different intervals, e.g. if we wrote

$$B : [0, 2] \times [-1, 10] \rightarrow [-5, 5],$$

then this would mean B is a function of two arguments; the first of which must be between 0 and 2; the second of which must be between -1 and 10; and the output of the function will be between -5 and 5. We can also leave the dimension of a function unfixed, e.g.

$$A : [0, 1]^n \rightarrow [0, 1].$$

Example 1.2. What will be the output of the function

$$f(x_1, x_2) = 2x_1 + 3x_2 - x_1^2$$

for the input vector $\langle -2, 5 \rangle$?

Solution. We substitute $x_1 = -2$ and $x_2 = 5$ into the function expression. This will give us

$$f(-2, 5) = 2(-2) + 3(5) - (-2)^2 = -4 + 15 - 4 = 7.$$

So our output is 7.

Example 1.3. Does the following medical test recommendation table define *Requires test* as a 'function' of *age*, *BMI* and *Family History*?

Age	BMI	Family history of condition	Requires test?
≤ 30	≤ 26	Yes	No
> 30	≤ 26	Yes	**Yes**
≤ 30	> 26	No	No
> 30	> 26	No	**Yes**

Solution. Although the table is not complete (e.g. it does not tell us what to do if the patient is above 30 with a BMI above 26 and a family history of the condition), it still satisfies the broad definition of a function since for each set of inputs, there is an unambiguous output. In this case it means there is a clear rule about whether a test is required (although it is only defined for these 4 combinations).

Example 1.4. Which of the following define multivariate functions on real numbers? If it makes sense, determine the output for the input vector $\mathbf{x} = \langle 2, -1, 3 \rangle$.

(i) The sum
(ii) The subtraction
(iii) The square root
(iv) The product
(v) Reordering from lowest to highest

Solution.

(i) The sum can be considered as a multivariate function (although 'addition' is usually thought of as a binary operation). The sum of 2, -1 and 3 is $2 + (-1) + 3 = 4$.
(ii) Although subtraction can be thought of as a multivariate function, which takes two values as its input, it is not defined for 3 values so we can't calculate "the subtraction of 3 numbers". We can subtract 3 numbers (from zero) or we could subtract the second two numbers from the first, but in general, subtraction or 'the difference' is something that mainly makes sense for two inputs. This is due to the property of associativity,

e.g. for addition $(a + b) + c = a + (b + c)$ however for subtraction $(a - b) - c \neq a - (b - c)$.

(iii) The square root is not a *multivariate* function as it only operates on one input.

(iv) The product can be considered as a multivariate function because it is always clear how the product of 2 or more arguments can be calculated. The product of 2, -1 and 3 is $2 \times (-1) \times 3 = -6$.

(v) Reordering could be considered as a function of a vector that produces a vector output, e.g. the input vector is $\langle 2, -1, 3 \rangle$ and the output vector is $\langle -1, 2, 3 \rangle$. However it is not the kind of multivariate function we will be looking at in this chapter, which take either multiple values (or equivalently vectors with length 2 or greater) and produce a singe value as their output. On the other hand, it can be thought of as a tuple of 3 independent functions called order statistics, which are special cases of the ordered weighted averaging operator we look at in Chap. 4.

1.3   The Arithmetic Mean

Although we don't usually think of it as such, the arithmetic mean is probably the multivariate function we most frequently come across. It is the "go to" operator when we want to summarize both small and large datasets. This is not just because it is quite simple to calculate. The arithmetic mean is also fairly unbiased (i.e. it is equally influenced by all of the inputs) and performs reasonably well for a wide range of prediction tasks. In some cases, it might not be the best choice if our data has more lower values than high values or vice versa, however if we don't know anything about our data, then it provides a good start for telling us about the central tendency.

1.3.1   Definition

Let us begin by defining the arithmetic mean, both informally and using a mathematical expression.

Definition (informal) 1.1 (Arithmetic Mean). The arithmetic mean (also referred to as 'the average' or 'mean') is the value obtained when we add up all our inputs and divide by how many there are. For example, if our inputs are 3, 7, and 8, then adding these gives 18, which when divided by 3 (since there are three numbers) gives 6, so our arithmetic mean would be 6.

Definition 1.1 (Arithmetic Mean). For an input vector $\mathbf{x} = \langle x_1, x_2, \ldots, x_n \rangle$, the arithmetic mean is given by

$$\text{AM}(\mathbf{x}) = \frac{1}{n} \sum_{i=1}^{n} x_i = \frac{x_1 + x_2 + \cdots + x_n}{n}$$

Notation Note Arithmetic mean

The fraction $\frac{1}{n}$ at the beginning is the same as dividing everything by n, i.e. multiplying by a half ($\frac{1}{2}$) means the same thing as dividing by 2.

The symbol $\sum$ is the Greek letter sigma and stands for 'sum'. It is used along with the index notation to indicate that we add up all of the x_i terms according to the indices specified by the limits above and below. The $i = 1$ below indicates the index we start at, while the n above is the index we finish at. So $\sum_{i=2}^{5} x_i$ would mean we take the sum $x_2 + x_3 + x_4 + x_5$. In practice, the value of n is implied by the number of arguments we have and we usually start at 1.

As previously mentioned, the notation x_i represents the i-th argument of our vector $\mathbf{x}$. So if $\mathbf{x} = \langle 2, 6, 7 \rangle$ then $x_2 = 6$ and there is no x_5 since our vector only has a length of 3.

When coming across any notation like this, it sometimes helps to explicitly write out the function for a few instances of n.

Example 1.5. Write out the arithmetic mean explicitly for 3 arguments.

Solution. For three arguments, we have $n = 3$ and so

$$\text{AM}(x_1, x_2, x_3) = \frac{1}{3}(x_1 + x_2 + x_3)$$

or

$$\text{AM}(x_1, x_2, x_3) = \frac{x_1 + x_2 + x_3}{3}$$

Example 1.6. Let $\mathbf{x} = \langle 3, 2, 9, 8 \rangle$.

  (i) What is the value of n?
 (ii) Calculate $\text{AM}(\mathbf{x})$.
(iii) What will be the output of $\text{AM}(\mathbf{x})$ if the order of the arguments is changed so that $\mathbf{x} = \langle 9, 8, 2, 3 \rangle$?

Solution.

  (i) In this case, $\mathbf{x}$ has 4 arguments and so $n = 4$.
 (ii) We know that $n = 4$, so we will have:

$$AM(3, 2, 9, 8) = \frac{1}{4} \sum_{i=1}^{4} x_i = \frac{1}{4}(3 + 2 + 9 + 8) = \frac{22}{4} = 5.5.$$

(iii) For $\mathbf{x} = \langle 9, 8, 2, 3 \rangle$, we have:

$$AM(9, 8, 2, 3) = \frac{1}{4}(9 + 8 + 2 + 3) = \frac{22}{4} = 5.5.$$

We see that even if the arguments are in a different order, we still obtain the same value. We will see that this is always a property of the arithmetic mean.

1.3.2 Properties

Of course we don't want to become lost in mathematical formalisms, however it is useful to understand some of the behaviors of the arithmetic mean so that we can understand its strengths, weaknesses, and when it is or isn't appropriate to use it (for more in-depth discussion concerning any of the following properties see [1, 2, 5, 8, 16] and the references therein.

Symmetry (The order of arguments doesn't affect the output). We already saw in the previous example that $AM(9, 8, 2, 3) = AM(3, 2, 9, 8)$. This kind of behavior holds for permuting vectors of any length. So if we know the value of $AM(1, 2.1, 3.5, 3, 9, 1)$ then we don't need to recalculate in order to determine the value of $AM(3.5, 3, 1, 2.1, 1, 9)$. Symmetry can also be interpreted as anonymity (we don't distinguish between the different sources of the inputs) or equal importance with respect to the arguments. Formally, we sometimes write

$$AM(\mathbf{x}) = AM(\mathbf{x}_\sigma), \text{ for all } \sigma$$

where σ indicates a permutation of the input vector.

Notation Note σ for permutations
The σ symbol is the Greek letter, small 'sigma' (as opposed to the big sigma $\sum$ we used for sum). Although it is used here in this way to indicate a general re-ordering of the vector, such notation is not universal (and you may have seen it used to indicate the standard deviation in statistics).

One permutation that we will make use of later is the arrangement of inputs into non-decreasing order (i.e. from lowest to highest). We will use $\mathbf{x}_\nearrow$ for vectors and $x_{(i)}$ for the arguments when arranged, so for $\mathbf{x} = \langle 3, 2, 9, 8 \rangle$, we have

$$\mathbf{x}_\nearrow = \langle x_{(1)}, x_{(2)}, x_{(3)}, x_{(4)} \rangle = \langle 2, 3, 8, 9 \rangle.$$

> **Side Note 1.1**   *The terminology "non-increasing" (descending) or "non-decreasing" (ascending) is usually adopted in more formal definitions rather than ascending or increasing, since equal values don't increase when ordered sequentially, i.e. the sequence 0.2, 0.3, 0.3 is not increasing but it is non-decreasing.*

Translation invariance (If a constant is added to every input and we take the average, the output will change by that same constant). For example, we have AM(4, 3, 8) = 5, and adding 3 to every input gives AM(7, 6, 11) = 8, so the output has also increased by 3. Formally we can write,

$$\mathrm{AM}(x_1 + t, x_2 + t, \ldots, x_n + t) = \mathrm{AM}(x_1, x_2, \ldots, x_n) + t.$$

Homogeneity (If you multiply every input by a constant, the output will change by the same factor). For example, AM(3, 7, 8) = 6, and multiplying every input by 5 results in AM(15, 35, 40) = 30. Formally this can be expressed,

$$\mathrm{AM}(\lambda x_1, \lambda x_2, \ldots, \lambda x_n) = \lambda(\mathrm{AM}(x_1, x_2, \ldots, x_n)).$$

Together, homogeneity and translation invariance tell us that the arithmetic mean is not dependent on the scale. Fahrenheit and Celsius for example are related by such a transformation where temperature in Fahrenheit is 1.8 multiplied by the temperature in Celsius added to 32. An *average of temperatures* in Celsius will therefore refer to the same temperature as the average of corresponding temperatures measured in Fahrenheit.

Monotonicity (Increasing one of the inputs does not result in a decrease to the output). For example, AM(40, 57, 30) > AM(40, 57, 29). Formally we can write,

$$\text{If } \mathbf{x} \leq \mathbf{y} \text{ then } \mathrm{AM}(\mathbf{x}) \leq \mathrm{AM}(\mathbf{y}).$$

Where the inequality between the vectors means that none of the arguments in **x** is greater than the corresponding arguments in **y**. In the case of the arithmetic mean, increasing one of the inputs in fact always results in an *increase* to the output.

Idempotency (If all the inputs are the same then the output will be the same too). For example AM(5, 5, 5, 5) = 5. This property is true for all functions that we refer to as 'averaging'. It is useful when we want the value of the output to resemble the inputs or be 'representative'.

> *Example 1.7.* Without calculating, which will be higher:
> AM(24, 68, 29, 39) or AM(31, 40, 24, 76)?
>
> *Solution.* Since the arithmetic mean is symmetric, we can reorder the arguments of the second mean calculation. We hence know that,

$$AM(31, 40, 24, 76) = AM(24, 76, 31, 40).$$

Comparing this reordered vector with the arguments in the first mean calculation, we have

$$24 = 24, \quad 68 < 76, \quad 29 < 31, \quad 39 < 40.$$

From monotonicity, we can therefore deduce that

$$AM(24, 68, 29, 39) < AM(31, 40, 24, 76).$$

Example 1.8. If $AM(29, 31, 40, 34, 11) = 29$, what will be the average if every output is multiplied by 10 and then has 51 added to it?
(i.e. $AM(341, 361, 451, 391, 161)$)

Solution. Since the arithmetic mean is homogeneous and translation invariant, we know that the average of these values will just be the original output multiplied by 10 with 51 added to it as well. $29 \times 10 = 290, 290 + 51 = 341$ so our average is 341.

Example 1.9. If the average height of 5 children is 94 cm and each grows 10 cm in a year, what will be the average height after 2 years?

Solution. We can again make use of the translation invariance property. The average height will also increase by 10 cm each year and so after 2 years the children will have an average height of 114 cm or 1.14 m.

1.4   The Median

Despite its usefulness and ease in calculation when summarising data, the arithmetic mean is not always the best choice.

If our aim is to produce an output that is 'representative', the arithmetic mean's susceptibility to skewed distributions and outliers has the practical implication that a single extreme value can distort the output. This is a standard criticism against any reporting methods that evaluate performance using the mean.

The **median** is one function that is used instead of the arithmetic mean in situations where we have reason to believe the data is not normally distributed or might contain 'outliers' (e.g. think of somebody making a typo when entering values into a spreadsheet so that 9 becomes 99). Housing prices and monthly incomes, for example, are reported in terms of the median selling price or median income rather than as means. The median is the middle number when the inputs are arranged in ascending or descending order.

> **Definition (informal) 1.2 (Median).**
> The median is the 'middle' number after we arrange our data in order (either ascending or descending) or, if we have an even number of numbers, the value half-way between the middle two numbers.

> **Definition 1.2 (Median).**
> For an input vector $\mathbf{x} = \langle x_1, x_2, \ldots, x_n \rangle$, the median is given by
>
> $$\text{Median}(\mathbf{x}) = \begin{cases} x_{(k)}, & n = 2k - 1 \\ \frac{x_{(k)} + x_{(k+1)}}{2}, & n = 2k. \end{cases}$$
>
> where k is a natural number, i.e. $1, 2, 3$, etc., and the index notation $(\cdot)$ means we first rearrange the data in order (here we use lowest to highest, although highest to lowest will produce same result).

> **Notation Note** Denoting even and odd with $n = 2k, n = 2k - 1$
> The cases $n = 2k$ and $n = 2k - 1$ refer to whether there is an even or odd number of data to aggregate. For example, if $n = 7$ then we would have $k = 4$ since $2 \times 4 - 1 = 7$ and hence the median is the fourth input after reordering. This notation used to denote a reordering of the inputs is not universally adopted, however it can be convenient generally for referring to the k-th highest value.

> *Example 1.10.* Let $\mathbf{x} = \langle 4, 6, 9, 1 \rangle$ and $\mathbf{y} = \langle 2, 6, 9, 4, 106 \rangle$. Calculate Median$(\mathbf{x})$ and Median$(\mathbf{y})$.
>
> *Solution.* We first arrange the data in ascending order. We will have
>
> $$\mathbf{x}_{\nearrow} = \langle 1, 4, 6, 9 \rangle$$
>
> This is an even number of values and so we take the simple average of the middle two. This gives $\frac{4+6}{2} = 5$. Then for $\mathbf{y}$, we have
>
> $$\mathbf{y}_{\nearrow} = \langle 2, 4, 6, 9, 106 \rangle$$
>
> In this case we have an odd number of values. The one in the middle is $y_{(3)} = 6$, and so this is our median.

The median satisfies all of the properties discussed for the arithmetic mean in the previous section, i.e. it is symmetric, translation invariant, homogeneous, idempotent and monotonic, however in the case of monotonicity we say it is not 'strictly' monotone because an increase to one of the values does not necessarily increase the output, although it still never results in a decrease.

1.5   The Geometric and Harmonic Means

The need for alternative measures of centre when there are outliers or skewed data is well known in introductory statistics courses, however there are other reasons why it sometimes just isn't appropriate to use the arithmetic mean to find the 'average'. Consider the following scenarios:

- Your pay goes up by 20 % one year and 10 % the next. What is the average pay increase?
- Leia can paint a house in 3 h, Luke can paint one in 5 h. How long would it take them to paint two houses if they work together?

1.5.1   The Geometric Mean

We will start with the first situation, which we will find should be aggregated using the geometric mean.

> **(Thinking Out Loud)**
> Income increasing each year is similar to our opening example. If our pay were $10,000 in the beginning, after a year we would have
>
> $$10,000 \times 1.2 = 12,000$$
>
> and the following year we would have
>
> $$12,000 \times 1.1 = 13,200.$$
>
> If we take the arithmetic mean of these two multipliers, we get 1.15, however from our starting value, over 2 years this would give
>
> $$10,000 \times 1.15 \times 1.15 = 13,225.$$
>
> So if we interpret the 'average' pay increase as the pay increase that, when applied both years, would produce the same result of $13,200, then what we are looking for is a value r such that
>
> $$10,000 \times r \times r = 10,000 \times 1.2 \times 1.1.$$

This leads us to a function known as the **geometric mean**.

Definition (informal) 1.3 (Geometric Mean).
For two or more arguments, the geometric mean is the value obtained when we multiply all of the inputs together and then take the n-th root (where n is how many values we have).

Definition 1.3 (Geometric Mean).
For an input vector $\mathbf{x} = \langle x_1, x_2, \ldots, x_n \rangle$, the geometric mean is given by

$$\mathrm{GM}(\mathbf{x}) = \left(\prod_{i=1}^{n} x_i \right)^{1/n} = (x_1 x_2 \cdots x_n)^{1/n} = \sqrt[n]{x_1 \times x_2 \times \cdots \times x_n}.$$

Notation Note Product $\prod$
The large $\prod$ symbol here is the Greek symbol pi and means product. The $i = 1$ at the bottom and n at the top are interpreted the same way they are for the sigma operator, so $\prod_{i=3}^{5} x_i$ would mean $x_3 \times x_4 \times x_5$. The fractional power $\frac{1}{n}$ is the same as saying the n-th root. The most common of these that you may have come across is the square root, so $a^{\frac{1}{2}}$ is the same as $\sqrt{a}$ and sometimes other numbers are written using the root sign along with the n value, e.g. $a^{\frac{1}{5}} = \sqrt[5]{a}$. A key thing to note is that when we multiply powers of the same base, the indices add, so $a^{\frac{1}{3}} \times a^{\frac{1}{3}} = a^{\frac{2}{3}}$. This means that if all our inputs are the same, in the index we will have $\frac{1}{n} + \frac{1}{n} + \ldots \frac{1}{n} = 1$, which gives us idempotency, but remember that we can't simplify like this if the bases are different, e.g. $a^{\frac{1}{2}} \times b^{\frac{1}{2}}$ does not simplify (although it can be written more simply is $\sqrt{ab}$).

The geometric mean should be used for averaging whenever our values are related by multiplication. In the case of two inputs, we can interpret the product as the area of a rectangle. The geometric mean then can be interpreted as the dimensions of a square that would be required to give the same area.

Example 1.11. Your pay goes up by 20 % one year and 10 % the next. What is the average pay increase?

Solution. Since our pay accumulates in terms of these increases and they are used for multiplication rather than addition, it is appropriate to use the geometric mean here. An increase of 20 % is the same as multiplying by 1.2, while an increase of 10 % is the same as multiplying by 1.1. We are therefore looking for the geometric mean of 1.1 and 1.2.

$$\mathrm{GM}(1.1, 1.2) = (1.1 \times 1.2)^{\frac{1}{2}} = \sqrt{1.32} \approx 1.1489.$$

Our average pay increase is approximately 14.89 %.

As mentioned in the notation note, the geometric mean satisfies idempotency, which means that it can still be interpreted as a representative value of the inputs. It is also symmetric and homogeneous, however it is not translation invariant, so we need to be careful if we apply any transformations to our data.

Side Note 1.2 *On this point it is worth mentioning that, for the arithmetic mean, the results will be equivalent whether we take the average of multipliers* $AM(1.2, 1.1) = 1.15$ *or the percentage increases* $AM(20\%, 10\%) = 15\%$ *(due to translation invariance and homogeneity). In the case of the geometric mean, however* $GM(1.2, 1.1) \neq GM(20, 10)$, *so we can't just take the geometric mean of the percentage increases (which would give a differing result of approximately 14.14%).*

The geometric mean will be strictly monotonic except if one of the inputs is zero. If any single input is zero, we will immediately obtain a zero result, even if we have thousands of inputs with high values. This leads us to some other aggregation properties worth noting.

1.5.2 More Properties

The following properties are relevant to the geometric mean and some other functions we will look at later on, however they are not usually discussed when using the arithmetic mean or median.

Absorbent elements An aggregation function is said to have an absorbent element a if including that value as *any* one of the inputs always results in an output of a. We can say that the geometric mean has an absorbent element of zero, or we can write

$$GM(x_1, x_2, \ldots, 0, \ldots, x_n) = 0$$

for 0 in any position.

Lipschitz continuity (or non-Lipschitz) The term Lipschitz continuity is not one we will have a strong focus on, although you might find it useful to be aware of. It means that the derivative or slope of the function (the change in the output as a fraction of the change in the input) is never infinitely large. A function that is 1-Lipschitz never has a gradient or slope larger than 1. The geometric mean is an example of a function that is *not* Lipschitz continuous. If one of the inputs is zero, then we have a zero output, however increasing this value to even a very small positive value can have a huge impact on the output. For example, suppose we have $\mathbf{x} = \langle 9293, 9568, 9460, 0 \rangle$. The geometric mean of these numbers is 0, but if we increase the 0 to 0.001 then we get approximately 170.3 (a rise over run of 170300). This kind of behavior can cause some trouble when we are optimizing, and in some applications where zero values are likely to occur, sometimes a small amount is added to each value. We have seen however that this is not technically allowed since the geometric mean is not translation invariant.

1.5.3 The Harmonic Mean

Let's return to our second scenario. Leia can paint a house in 3 h, Luke can paint one in 5, and we want to know how long it would take them to paint two houses if they work together. This is equivalent to asking how long it would take two people who both worked at Leia and Luke's 'average' pace.

(Thinking Out Loud)

If we surmized that the average pace should be 1 house in 4 h (the arithmetic mean of 3 and 5), thinking things through we would see that in this time, Leia can paint $1\frac{1}{3}$ houses, while Luke would have painted $\frac{4}{5}$. This would put us *over* two houses.

The problem is that the rates are given to us in a somewhat reciprocal fashion. Usually we give rates with respect to time, so we have:

  Leia paints a house at the rate of $\frac{1}{3}$ of a house per hour.

  Luke paints at the rate of $\frac{1}{5}$ of a house per hour.

From this, we can work out how many houses they would paint together per hour by adding the values.

$$\frac{1}{3} + \frac{1}{5} = \frac{5}{15} + \frac{3}{15} = \frac{8}{15}.$$

Now that we know this rate, we can ask how long it takes to paint 2 houses, which is

$$2 \div \frac{8}{15} = 2 \times \frac{15}{8} = \frac{30}{8} = 3.75.$$

This kind of problem is often posed although the scenario can be different. In fact, it is similar to asking the question 'If a car travels at 100 km/h on the way to its destination and returns at a speed of 80 km/h, what is its average speed?' The key aspect in both situations is that the way we combine the numbers to find the 'average' is reciprocal to the way in which they are given or their units.

In asking how long it takes for the painters to finish the houses, we need to know their rate of houses painted with respect to time, but we are given the time with respect to houses.

On the other hand, for a car traveling different speeds over the same distance, the average speed is the total distance divided by the total time, however taking the simple arithmetic mean of 100 and 80 does not take into account that the car travels at these speeds for *different amounts of time*.

The correct average to use in these situations is the harmonic mean.

Definition (informal) 1.4 (Harmonic Mean).
To find the harmonic mean of a set of numbers, we find the average of their reciprocal values (i.e. in fraction form this means we flip the number upside down) and then take the reciprocal of this result. For example, if we are taking the harmonic mean of 4, 3 and 2, we find the average of $\frac{1}{4}$, $\frac{1}{3}$ and $\frac{1}{2}$, which gives $\frac{13}{12} \div 3 = \frac{13}{36}$ and taking the reciprocal of this gives $\frac{36}{13}$ which is approximately 2.769.

Definition 1.4 (Harmonic Mean).
For an input vector $\mathbf{x} = \langle x_1, x_2, \ldots, x_n \rangle$, the harmonic mean is given by

$$HM(\mathbf{x}) = n \left(\sum_{i=1}^{n} \frac{1}{x_i} \right)^{-1} = \frac{n}{\frac{1}{x_1} + \frac{1}{x_2} + \cdots + \frac{1}{x_n}}.$$

If any of the inputs is 0, we define the HM to give an output of 0.

Notation Note Negative powers
The only unfamiliar symbol we see here is the -1 power. Generally, a negative power $-p$ means that we divide 1 by the term raised to the power of p. So 3^{-1} means $\frac{1}{3^1} = \frac{1}{3}$ and 3^{-2} means $\frac{1}{3^2} = \frac{1}{9}$. Here it just means that we divide 1 by the sum of everything in the brackets.

Example 1.12. Leia can paint a house in 3 h, Luke can paint one in 5 h. How long would it take someone who works at the average of their pace?

Solution. This is an average that is appropriate to calculate using the harmonic mean. We have

$$HM(3,5) = 2 \left(\frac{1}{\frac{1}{3} + \frac{1}{5}} \right) = 2 \left(\frac{1}{\frac{5}{15} + \frac{3}{15}} \right) = \frac{2}{\left(\frac{8}{15} \right)} = \frac{30}{8} = 3.75.$$

Like the geometric mean, the harmonic mean is symmetric, idempotent, homogenous and has an absorbent element of 0. It also is *not* translation invariant.
Unlike the geometric mean, however, the harmonic mean *is* Lipschitz continuous. Although any input being zero will make the function equal to zero, the rate of increase near these values is not as drastic as the geometric mean.

Example 1.13. Calculate the arithmetic mean, geometric mean and harmonic mean for the input vector $\mathbf{x} = \langle 0.6, 0.9, 0.26, 0.7 \rangle$.

Solution. The arithmetic mean will be

$$AM(0.6, 0.9, 0.26, 0.7) = \frac{1}{4}(0.6 + 0.9 + 0.26 + 0.7) = 0.615.$$

For the geometric mean, we have

$$GM(0.6, 0.9, 0.26, 0.7) = (0.6 \times 0.9 \times 0.26 \times 0.7)^{1/4} \approx 0.560.$$

For the harmonic mean, the value will be

$$HM(0.6, 0.9, 0.26, 0.7) = \frac{4}{\frac{1}{0.6} + \frac{1}{0.9} + \frac{1}{0.26} + \frac{1}{0.7}} \approx 0.497.$$

An interesting well known result about these means is that for any input vector $\mathbf{x}$, it always holds that

$$HM(\mathbf{x}) \leq GM(\mathbf{x}) \leq AM(\mathbf{x}).$$

Plots[2] to compare each of the functions are shown in Fig. 1.1.

1.6 Averaging Aggregation Functions

We have used the term 'mean' for the arithmetic mean, the geometric mean, and the harmonic mean. We have also referred to these and the median as 'averages'. Usually when people say 'average' and 'mean' they are referring to the arithmetic mean, however the class of averaging functions is actually much broader.

Although it is not universally adopted terminology, **averaging behavior** for real numbers can be described in terms of the output being some number in between the minimum and maximum inputs. With this property, we can still say that such functions in some way give an output that is 'representative' of the inputs and that can be interpreted in the same units and scale.

For similar reasons, sometimes any function that is idempotent is considered to be 'averaging', although these two properties are not equivalent.

> **Side Note 1.3** *However any function that is averaging will necessarily be idempotent and any function that is both idempotent and monotonic will be averaging.*

The term *aggregation function* is also sometimes synonymous with averaging functions, however we will distinguish the latter as a special case of the former by using the following definitions.

---

[2]Generated using the `f.plot3d()` function from the `AggWAfit` library available from http://www.researchgate.net/publication/306099814_AggWAfit_R_library or alternatively http://aggregationfunctions.wordpress.com/book.

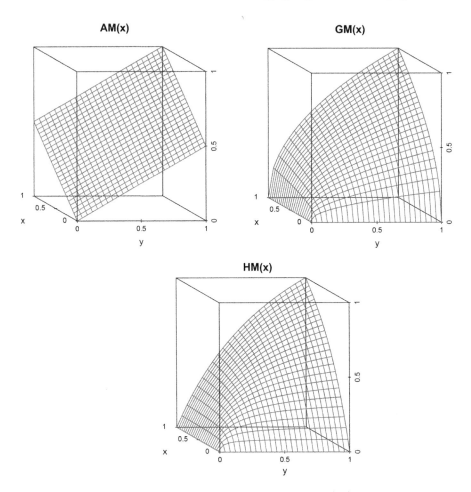

Fig. 1.1 Plots of 2-variate instances of the arithmetic mean $AM(x, y)$, the geometric mean $GM(x, y)$ and the harmonic mean $HM(x, y)$

Definition 1.5 (Aggregation Function).
For an input vector $\mathbf{x} = \langle x_1, x_2, \ldots, x_n \rangle$, a multivariate function $A(\mathbf{x})$ can be referred to as an aggregation function if it:

- is monotone increasing (in either a strict or non-strict sense), i.e. an increase to any of the inputs cannot result in a decrease to the output. We can express this property as:

$$\text{If } \mathbf{x} \leq \mathbf{y} \text{ then } A(\mathbf{x}) \leq A(\mathbf{y});$$

- satisfies the boundary conditions

$$A(a, a, \ldots, a) = a, \quad A(b, b, \ldots, b) = b,$$

where a and b are the minimum and maximum values possible.

For example, with the boundary conditions, if all our inputs take a value over the unit interval $[0, 1]$ then we have $A(0, 0, \ldots, 0) = 0$ and $A(1, 1, \ldots, 1) = 1$. If our data range is from 1 to 1000, then we would require $A(1, 1, \ldots, 1) = 1$, $A(1000, 1000, \ldots, 1000) = 1000$. However we do not require idempotency in general in order for something to be an aggregation function. A function like the product $A(x, y) = xy$ is an aggregation function over the interval $[0, 1]$ because it is monotone and $f(0, 0) = 0$ and $f(1, 1) = 1$, but not if our interval is $[1, 10]$ since $10 \times 10 = 100$.

We then consider averaging functions to be a special sub-class of aggregation functions.

Definition 1.6. An aggregation function $A(\mathbf{x})$ is averaging if its output is bounded between the minimum and maximum of its inputs, i.e.

$$\min(\mathbf{x}) \le A(\mathbf{x}) \le \max(\mathbf{x})$$

Example 1.14. Show that the 3-variate function

$$f(x_1, x_2, x_3) = \frac{x_1}{2}(x_2 + x_3)$$

is an aggregation function for $x_1, x_2, x_3 \in [0, 1]$.

Solution. First we can check the boundary conditions. It holds that $f(0, 0, 0) = 0$ (since if x_1 is zero then the output will be zero) and $f(1, 1, 1) = \frac{1}{2}(1 + 1) = 1$. By looking at the structure of the function, we can also see that if any of x_1, x_2, x_3 increases, then the output cannot decrease since both multiplication and addition are monotone operations.

Notation Note Element of a set $\in$
The symbol $\in$ is read as "is an element of". Specifying that $x \in [0, 1]$ is the same as $0 \le x \le 1$. For vectors, we can also write $\mathbf{x} \in [0, 1]^3$ which means $\mathbf{x}$ is a vector of three arguments, all of which are between 0 and 1.

1.7 Uses of Aggregation Functions

The arithmetic mean is used almost everywhere when the need to summarize data arises. A few interesting examples are provided below.

1.7.1   Sporting Statistics

The arithmetic mean is used in sports to give an idea of a player's ability or performance, for example:

Points per game in basketball—In her rookie season, Elena Delle Donne of the Chicago Sky had an average of 18.1 points per game, scoring 543 points out of her 30 regular season games played [17];

Batting average in baseball—In his rookie year, Hayato Sakomoto of the Tokyo Giants had a batting average of 0.333 since he scored 1 hit out of three times at base [13], i.e. $(1 + 0 + 0)/3$. This kind of statistic is interesting because for each time at base it is only possible to either score a hit or not. It's impossible for Mr. Sakomoto to ever get 0.333 hits when he's at base, however, if he sustained this average over more innings we might develop the expectation that out of ten times at base he might score a hit about three times.

There are also examples of statistics in sport whose 'average' would require the harmonic mean rather than the arithmetic.

Bowling strike rate in cricket—Curtly Ambrose had a career test strike rate of 54.6, having taken 405 wickets in 22,103 balls [10]. The natural expression of a strike rate as *balls per wicket*, rather than *wickets per ball*, is similar to describing our painters' rates in *hours per house*. If we had two bowlers, each having bowled the same number of balls, the average of their strike rates should be evaluated using the harmonic mean—since this would be equivalent to their total number of balls bowled divided by the total number of wickets taken.

1.7.2   Simple Forecasting

Daily sales—There is a lot of uncertainty when it comes to restaurants being able to predict daily sales, with weekly, yearly and seasonal trends to take into account. One simple method used is to adjust the figures by looking at the average of the same day over the last few weeks.

Central estimates—Insurance companies plan their pricing (at least partly) based on evaluations of expected compensation losses. Independent estimates can be collected using different models and methods, after which, if they are considered equally reasonable, these can be averaged using the arithmetic mean to obtain a 'central' estimate [12].

1.7.3   Indices of Diversity and Equity

A number of indices are employed in ecological management and research to quantify levels of biodiversity. Many of these indices were derived from models used in economics.

The Gini index of wealth inequality—For a set of incomes, the Gini index [6] calculates the arithmetic mean of the pairwise differences between the individual incomes. For example, if we had four individuals and their incomes were given by $\mathbf{x} = \langle 3, 6, 8, 20 \rangle$ then the Gini index would be the average of $6 - 3 = 3, 8 - 3 = 5, 20 - 3 = 17, 8 - 6 = 2, 20 - 6 = 14$ and $20 - 8 = 12$ to give $\frac{53}{6} \approx 8.8$. This is usually normalized to a value between 0 and 1. A similar calculation is used in ecology [3] where, instead of incomes, different species abundances are used.

Poverty indices—There are various indices other than income-based evaluations (e.g. [4]) which essentially take the arithmetic mean of scores over a number of criteria.

Geometric mean of species abundances—Researchers in ecology have recently used the geometric mean of species abundances as a proxy for biodiversity [9, 11]. This is an interesting case because the reason the geometric mean is chosen is not because the species abundances would usually be combined using multiplication as such, but rather due to the following properties:

- The output of the geometric mean gets very small if any of the inputs are close to zero. The interpretation here is that we want the function to be more sensitive to increases in rare species;
- As we will see in Chap. 2 (Sect. 2.9), the geometric mean can be associated with the sum of log transformations of the inputs. A log transformation is useful in situations where there may be some infrequent but very high values. For example, if we are looking at incomes at a large scale company, most of the employees will have low to moderate incomes, while the CEO might be earning millions. Similarly in ecology, there may be some very common species, with populations substantially higher than the next most frequently occurring. By using a geometric mean, we limit the ability of a single very high value to bring up the entire average.

1.8   Summary of Formulas

Arithmetic Mean

$$\text{AM}(\mathbf{x}) = \frac{1}{n} \sum_{i=1}^{n} x_i = \frac{x_1 + x_2 + \cdots + x_n}{n} \tag{1.1}$$

Geometric Mean

$$GM(\mathbf{x}) = \left(\prod_{i=1}^{n} x_i\right)^{1/n} = (x_1 x_2 \cdots x_n)^{1/n} \tag{1.2}$$

Harmonic Mean

$$HM(\mathbf{x}) = n\left(\sum_{i=1}^{n} \frac{1}{x_i}\right)^{-1} = \frac{n}{\frac{1}{x_1} + \frac{1}{x_2} + \cdots + \frac{1}{x_n}} \tag{1.3}$$

Aggregation Functions

For $\mathbf{x} \in [a, b]^n$, $A(\mathbf{x})$ is an aggregation function if it satisfies

$$A(a, a, \ldots, a) = a$$
$$A(b, b, \ldots, b) = b \text{ (boundary conditions)} \tag{1.4}$$

$$\text{If } \mathbf{x} \leq \mathbf{y} \text{ then } A(\mathbf{x}) \leq A(\mathbf{y}) \quad \text{(monotonicity)}$$

Averaging Functions

An aggregation function $A(\mathbf{x})$ defined over $[a, b]^n$ is averaging if

$$\min(\mathbf{x}) \leq A(\mathbf{x}) \leq \max(\mathbf{x}), \text{ for all } \mathbf{x} \in [a, b]^n \tag{1.5}$$

1.9   Practice Questions

1. Write down the arithmetic mean, the geometric mean and the harmonic mean explicitly for 3 and 4 arguments (i.e. in terms of x_1, x_2, x_3, x_4).
2. Using the concept of monotonicity, give some reasoning as to why it always holds that $G(\mathbf{x}) \geq \min(\mathbf{x})$. [Hint: remember that idempotency of the geometric means we will have $GM(a, a, \ldots, a) = a$ when all the arguments are the same]
3. Are the harmonic mean and geometric mean homogeneous and translation invariant? How about the median?
4. If $AM(23, 42, 27, 68) = 40$, without calculating directly, what will be the value of

$$AM(29, 48, 33, 74)?$$

5. If $AM(x_1, x_2, x_3, x_4) = 15$, what will be the value of

$$AM(x_1 + 2, x_2 + 2, x_3 + 2, x_4 + 2)?$$

6. If $AM(x_1, x_2, x_3, x_4) = 341$, what will be the value of

$$AM(x_1 + c, x_2 + c, x_3 + c, x_4 + c)?$$

7. For the vector $\mathbf{x} = \langle 11, 21, 34, 30 \rangle$ the arithmetic mean is 24, the geometric mean is 22.03196, and the harmonic mean is 19.87348, what will be the calculations for these functions if $\mathbf{x} = \langle 110, 210, 340, 300 \rangle$?

8. If $GM(x_1, x_2, x_3, x_4) = 6$, what will be the value of

$$GM(3x_1, 3x_2, 3x_3, 3x_4)?$$

9. If $AM(x_1, x_2, x_3, x_4) = 7$, what will be the value of

$$AM(3x_1 + 2, 3x_2 + 2, 3x_3 + 2, 3x_4 + 2)?$$

10. Which of the functions we've introduced so far are Lipschitz continuous?
11. Explain what it means to say that the arithmetic mean, harmonic mean and geometric mean could be affected by outliers.

1.10   R Tutorial

There are many online tutorials available for becoming familiar with the programming language R [15]. Once you are familiar with the basic commands, you will find that an online search for how to do something specific can almost always find an existing solution either as part of standard packages, or as code that someone else has provided. We will start with some basic operations and then move toward being able to calculate values for the functions we've introduced in this first chapter.

1.10.1   Entering Commands in the Console

For simple one-line commands, we can enter these straight into the console by typing the command and pressing return/enter.

If we want to retrieve a previous command, we can press the up arrow.

1.10.2   Basic Mathematical Operations

Elementary functions and basic arithmetic operations are all part of R. In most cases, these are the same as what you may have used before in programs like Microsoft Excel.

R Exercise 1 *Try entering in the following commands. Type what is written in the second 'in R' column and press enter to see if you obtain the expected result.*

Operation	Symbol	in R	Example	in R	Expected result
Addition	$+$	+	$5 + 2$	5+2	7
Subtraction	$-$	-	$6 - 8$	6-8	-2
Multiplication	$\times$	*	3×4	3*4	12
Division	$\div$	/	$3 \div 4$	3/4	0.75
Powers	$(\cdot)^p$	^ or **	3^4	3^4	81

1.10.3 Assignment of Variables

We are often going to be working with general operations in terms of variables and in some cases we will require operations on vectors.

We can use almost any word or letter for a variable along with the full stop symbol " . " and underscore " _ ". We can't use the dash (because it is treated as a minus sign) and we can't start with a number. There are also some reserved keywords, for example 'for', 'function', 'while', etc., that we can't use because they are already used by R for other things (we can check this by typing the function and an open bracket, e.g. "max (" and seeing whether the function parameters come up in the bottom status bar.

To assign a value to a variable we use the *less than* symbol '<' and the dash '-' together (it makes an arrow). So if we want to set the value of a variable a to 3, we write

```
a <- 3
```

Try doing this and see what happens when you enter the command a+4.

R Exercise 2 *Assign the following values to each variable.*

```
a <- 3
b <- 7
the.value <- 12
the.index <- 2
the.index_2 <- 3
```

Now evaluate:

```
a*b
the.value * a
a^the.index + b^the.index_2
```

1.10.4 Assigning a Vector to a Variable

In order to assign a vector to a value, we use a 'c' (which stands for 'combine' or 'concatenate') along with the entries of the vector inside normal brackets, separated by commas. For example,

```
a <- c(3,2,1,0)
```

We have some other special commands that we can also use. If we want a vector of a set of sequential numbers, e.g. ⟨4, 5, 6, 7, 8⟩ we can write the starting and finishing numbers separated by a colon.

```
a <- 4:8
```

We can also use `seq()` for this purpose. Using `seq(4)` will produce the sequence from 1 to 4, `seq(3,8)` will produce the sequence from 3 to 8, and we can also count in multiples, e.g. `seq(4,20,5)` produces the sequence from 4 to 20 counting by 5s, i.e. 4, 9, 14, 19. Sometimes we will need to first define an empty vector, or a vector with default starting entries. This is more easily achieved using the `array()` command in R. This command has two arguments. The first one is the default entry of the array, e.g. 0, while the second argument gives the length. For example,

```
a <- array(0,3)
b <- array(1,10)
long.array <- array(2,200)
```

We can also assign arrays that have a small sub-sequence. For example, suppose we want the vector ⟨1, 2, 3, 1, 2, 3, 1, 2, 3⟩. We can assign this using a combination of the `array()` and `c()` commands.

```
my.seq <- array(c(1,2,3),9)
```

Notice that in this case, the `c(1,2,3)` acts as just one argument in the array command, even though it's a set of 3 numbers.

The array command can also be used for storage of larger objects like tables. Each of the `array()` commands above can be achieved with `rep()`, e.g. `rep(1, 10)` produces a vector of length 10 that are all 1s, `rep(1:2, 10)` produces a repetition of 10 times the sequence 1,2 and `rep(1:2, length.out=9)` produces a vector of length 9 by repeating the sequence 1,2.

1.10.5 Basic Operations on Vectors

All of the basic operations we discussed, i.e. `+`, `-`, `*`, `/`, `^`, can be used on vectors. If we are operating with a vector and a normal number value (e.g. 3) these are calculated component-wise.

> **R Exercise 3** *Enter the following commands from the input column and check the results.*
>
> | Input | Expected output |
> |---|---|
> | `c(1,2,3) + 3` | *4 5 6* |
> | `c(1,2,3)*4` | *4 8 12* |
> | `c(1,2,3)^2` | *1 4 9* |

We can also have operations between vectors. In most cases, these operations should be between vectors of the same length. The operation is applied between each of the corresponding components, for example,

$$\langle 2,3 \rangle + \langle 5,-1 \rangle = \langle 2+5, 3+(-1) \rangle = \langle 7,2 \rangle.$$

> **R Exercise 4** *Assign the following two vectors,*
>
> ```
> a <- c(1,6,7,9)
> b <- c(-1, 2, 1, -2)
> ```
>
> *Now check the following commands against the expected output.*
>
> | Input | Expected output |
> |---|---|
> | `a+b` | *0 8 8 7* |
> | `a - b` | *2 4 6 11* |
> | `a*b` | *-1 12 7 -18* |
> | `a/b` | *-1 3 7 -4.5* |
> | `a^b` | *1 36 7 0.01234568* |

1.10.6 Existing Functions

The mean and the median are already pre-programmed into R. We can calculate the arithmetic mean of a set of numbers using `mean()` where the argument is a vector. For example,

```
mean(c(2,3,6,7))
mean(2:7)
mean(my.seq)
mean(c(4,a,1:3,the.value,long.array))
```

Notice that in this last example, we could combine numbers, pre-assigned values and vectors to use as the input. We just need to remember to use `c()` and separate the values by a comma. The `mean()` function has some other optional parameters but we will not use these. If we entered something like `mean(3,2,51,2)`, the value returned would only be based on the first number, i.e. 3, so we need to be careful. However sum(1,2,3) produces the output of 6.

The median function is also pre-programmed into R.

```
median(1:6)
median(c(1,7,82,2))
median(my.seq)
```

A few other pre-programmed functions that will be useful for us are sum(), prod() and length() which return the sum of the entries of the vector, the product of the values in the vector, and the length of the vector respectively.

We also have min() and max() functions, which can take multiple arguments including a combination of single values or vectors and return the minimum and maximum arguments.

These can also be combined with the previous operations we looked at, so sum(a^2) would first square every value in a, and then add all these values together.

R Exercise 5 *For the following vectors,*

```
a <- c(1,8,3,9)
b <- c(2,2,1,1)
d <- c(3,4,6,81,9)
```

Evaluate:

Input	Expected output
sum(a)	*21*
prod(c)	*52,488*
length(b)	*4*
sum(a*b)	*30*
sum(a^b)	*77*
prod(a)^(1/length(b))	*3.833659*
min(d)	*3*
max(a,d)	*81*
min(max(a),max(b))	*2*

Side Note 1.4 *Make sure you are careful with your brackets! The commands follow order of operations, which means without brackets in something like* prod(a)^(1/length(b)), *for example if you just entered* prod(a)^1/length(b), *then we would have the product to the power of 1 (which is just the product), which is then divided by its length, so it would be* $(1 \times 8 \times 3 \times 9)/4 = 54$ *and not the geometric mean* $(1 \times 8 \times 3 \times 9)^{1/4}$.

1.10.7   Creating New Functions

In many situations it is convenient for us to be able to program our own functions. This is actually relatively easy to do in R and can become a very powerful too. Creation of a basic function essentially consists of three components:

- Pre-defining the function inputs;
- A sequence of calculations;
- Return of an output.

We can start with an example that calculates the arithmetic mean (even though we don't need it because it is already defined in R with mean()).

```
our.mean <- function(x) {
sum(x)/length(x)
}
```

The value returned (i.e. the output) is always the calculation on the last line of our function before the } bracket. In this case, our whole calculation fits on one line, so we don't need to do any temporary assignment of values throughout. However we could if we wanted to. Here is another example that will return the same output.

```
our.mean.2 <- function(x) {
n <- length(x)
s <- sum(x)
output <- s/n
output
}
```

In both of these examples, the x input is expected to be a vector. We could now calculate the mean using these R functions, e.g. by entering our.mean(a) or our.mean(c(38,27,1)). You will notice that the variables n and s are not saved in the R workspace (i.e. if you type 'n' and press enter, R will say that 'n' is not found. They are referred to as 'local' variables.

If we only ever wanted to take the arithmetic mean of two arguments, we could program the function in terms of two inputs where each input is expected to be a number (rather than a vector).

```
mean.two <- function(x,y) {
(x+y)/2 }
```

In this case the average of 5 and 8 would be calculated using mean.two(5,8) rather than requiring c(5,8). However in this case, mean.two(5,8,13) won't return a value, and the output if x and y are vectors (of the same length) will be a vector as well (the component-wise arithmetic means).

If we do an assignment on the last line of the function, no output will be printed in the console, e.g.,

```
nothing.happens <- function(x) {
sum(x) +2
a <- prod(x)
}
```

However, R will still make these calculations when you use the function, and we can assign values with the output, e.g. `out <- nothing.happens(c(2,3))` will assign a 6 to `out`.

Let's now program functions for our geometric and harmonic means.

```
GM <- function(x) {
prod(x)^(1/length(x))
}


HM <- function(x) {
length(x)/sum(1/x)
}
```

Side Note 1.5 *When functions have just one line of calculation, there is no need for us to set them out over multiple lines like this, and we can omit the {...} brackets, i.e. if we wanted to, we could enter the harmonic mean using:*

```
HM <- function(x)   length(x)/sum(1/x)
```

and the function would be the same. In some future functions, however, we will need several lines of calculation. When entering multi-line functions in R, you can either do this directly in the command console (a '+' symbol will appear to let you know that you haven't closed off the necessary brackets), or you can hold the function in a library file and then copy and paste the whole sequence to the command console. In support applications like RStudio there are other ways to execute the code.

1.11   Practice Questions Using R

1. Suppose you have $\mathbf{x} = \langle 0.1, 0.2, 0.6, 0.9 \rangle$, Calculate

   (i)   The arithmetic mean
   (ii) The geometric mean
   (iii) The harmonic mean
   (iv) The median and compare the results.

2. Create 2-variate versions of the geometric mean and harmonic mean using `F <- function(x,y) {...}`, i.e. where the x and y are two numbers rather than vectors.

3. Define the function $f(x,y) = \frac{x^2+y^2}{x+y}$, $(x + y \neq 0$ and $f(0,0) = 0)$ and evaluate $f(0.3, 0.9)$, and $f(0.4, 0.9)$. Based on your results, can it be stated that f is *not* an aggregation function?

4. The rise in house prices in Australia's 8 major cities was 6.8 % over 2014 and 9.8 % over 2013. What is the average yearly increase? (which operator should be used?)

5. A car travels at 110 km/h on the way to a destination and 80 km/h on the way back. What is its average speed? (which operator should be used?)

6. Let $\mathbf{x} = \langle 25, 14, 39, 21, 51, 22 \rangle$. Compare the outputs of the arithmetic, harmonic, geometric means and the median. How do these values differ if the last input $x_6 = 22$ is replaced with an outlier $x_6 = 288$?

7. Let $\mathbf{x} = \langle 189, 177, 189, 212, 175, 231 \rangle$. Compare the outputs of the arithmetic, harmonic, geometric means and the median. How do these values differ if the last input $x_6 = 231$ is replaced with an outlier $x_6 = 11$?

References

1. Beliakov, G., Pradera, A., Calvo, T.: Aggregation Functions: A Guide for Practitioners. Springer, Heidelberg (2007)
2. Beliakov, G., Bustince, H., Calvo, T.: A Practical Guide to Averaging Functions. Springer, Berlin/New York (2015)
3. Camargo, J.A.: Must dominance increase with the number of subordinate species in competitive interactions? J. Theor. Biol. **161**(4), 537–542 (1993)
4. Economist Intelligence Unit: Women's economic opportunity 2012: A global index and ranking from the Economist Intelligence Unit, 1–51. http://www.eiu.com (2015). Cited 10 Aug 2015
5. Gagolewski, M.: Data Fusion. Theory, Methods and Applications. Institute of Computer Science, Polish Academy of Sciences (2015)
6. Gini, C.: Variabilità e Mutabilità. Tipografia di Paolo Cuppini, Bologna (1912)
7. Goldberg, D.: What every computer scientist should know about floating-point arithmetic. ACM Comput. Surv. **23**(1), 5–48 (1991)
8. Grabisch, M., Marichal, J.-L., Mesiar, R., Pap, E.: Aggregation Functions. Cambridge University press, Cambridge (2009)
9. Hale, S., Nimmo, D.G., James, S., White, J., et al.: Fire and climatic extremes shape mammal distributions in a fire-prone landscape. Diversity and Distributions. doi:10.1111/ddi.12471
10. Howstat Computing Services: Curtley Ambrose Player Profile - Test Cricket. http://www. howstat.com.au/cricket/statistics/Players/PlayerOverview.asp?PlayerId=0065 (2016). Cited 15 Jan 1999
11. Kelly, L.T., Bennett, A.F., Clarke, M.F., McCarthy, M.A.: Optimal fire histories for biodiversity conservation. Conserv. Biol. **29**, 473–481 (2015)
12. Lurie, P.: Actuarial Methods in Health Insurance Provisioning, Pricing and Forecasting. Institute of Actuaries of Australia Biennial Convention 2007, Christchurch, New Zealand (2007)
13. Nippon Professional Baseball Organisation (NPB): Hayato Sakomoto Player Statistics. http:// npb.jp/bis/eng/players/51955114.html (2016). Cited 15 Jul 2016
14. Ponte, J.P.: The History of the Concept of Function and Some Educational Implications. Math. Edu. **3**(2), 3–8 (1992)

15. R Core Team: R: A language and environment for statistical computing. R Foundation for Statistical Computing, Vienna, Austria. http://www.R-project.org/ (2014)
16. Torra, V., Narukawa, Y.: Modeling Decisions. Information Fusion and Aggregation Operators. Springer, Berlin/Heidelberg (2007)
17. Women's National Basketball Association (WNBA): Elena Delle Donne Player Statistics. http://www.wnba.com/player/elena-delle-donne/#/stats (2016). Cited 15 Jul 2016

Chapter 2
Transforming Data

Now we know some basics about summarizing data, however sometimes the data isn't always in the right format. Height, body mass, and points scored per game might be useful to combine together in order to summarize the appeal of a basketball player, but height could be given in centimeters or inches, body mass in kilograms or pounds, and these quantities *vary* in different ways too, so putting them together, e.g. by taking their arithmetic mean, might not give sensible results.

Are we still allowed to aggregate? Of course we are! We just have to make some wise decisions about how best to transform this data so that the output is useful. In this chapter, we will look at some alternative ways to transform data and then introduce the power means, a general family of means that includes the means we looked at in the previous chapter as special cases.

Assumed Background Concepts

- Standard deviation
  What would you expect to be the standard deviation for heights? What is the sample standard deviation for 3, 5, 8, 2, 3?

- Log and exponential arithmetic
  What does $\ln 3 + \ln 4$ simplify to? What is the value of $e^x(e^{\frac{1}{2}x} - e^{\frac{1}{3}x})$?

- Linear functions
  Can you express the straight line that runs between the points $(2, 3)$ and $(-1, 8)$? Can you draw the line representing the function $y = 5x - 2$ over the domain $[4, 7]$?

- Inverse functions
  What is the inverse function for x^2?

Chapter Objectives

- Understand the different roles that transformation of variables can play in pre- and post-processing of data
- To introduce the *power means*, which generalize the geometric, harmonic and arithmetic means
- To build intuition about using transformations appropriately

---

© Springer International Publishing AG 2016

S. James, *An Introduction to Data Analysis using Aggregation Functions in R*,
DOI 10.1007/978-3-319-46762-7_2

2.1   The Problem in Data: Multicriteria Evaluation

Let's start with an exercise (inspired by Lesh et al. [7]). Suppose we are charged with splitting the following 20 students[1] into two "fair" volleyball teams.

Student	Sprint 100 m (seconds)	Height (cm)	Serving (out of 100)	Endurance (out of 30)
Mizuho	15.78	148	94	17
Yukie	21.15	147	94	20
Megumi	14.30	134	91	17
Sakura	19.59	174	88	16
Izumi	10.96	145	93	16
Yukiko	19.17	158	83	12
Yumiko	18.35	157	99	20
Kayoko	14.09	177	82	23
Yuko	27.98	155	93	19
Hirono	16.51	165	85	7
Mitsuko	15.57	137	100	14
Haruka	14.16	162	93	16
Takako	22.40	176	95	15
Mayumi	21.34	153	97	9
Noriko	15.67	140	94	8
Yuka	19.12	155	81	3
Satomi	21.50	147	88	5
Fumiyo	40.29	161	95	19
Chisato	12.34	160	89	26
Kaori	13.38	134	81	16

We could set about this task in a few ways. One approach would be to base our decision on just one of the variables, however obviously this would have the drawback that even if we had two teams that were fairly matched in height, we might end up having all of the fast girls on one team and the slow ones on the other.

If we can get an 'overall rating' or ranking of each player, we could then use this overall statistic to split our teams, and one way to determine this rating is by aggregating the variables. However, looking closely at the data we come across a few problems.

Let's just focus on the first 8 girls in the list. Taking the average of each variable would give the following.

---

[1]The female student names here are borrowed respectfully from Kōshun Takami's novel *Battle Royale* (although his novel is not related at all to volleyball).

Student	Sprint	Height	Serving	Endurance	AM
Mizuho	15.78	148	94	17	68.55
Yukie	21.15	147	94	20	70.43
Megumi	14.30	134	91	17	64.36
Sakura	19.59	174	88	16	74.50
Izumi	10.96	145	93	16	66.37
Yukiko	19.17	158	83	12	68.06
Yumiko	18.35	157	99	20	73.44
Kayoko	14.09	177	82	23	73.92

2.1.1   Which is Better: Higher or Lower?

One thing that you might notice immediately is that a slower sprinting time contributes to a higher value in the arithmetic mean. Mizuho is a lot slower than Izumi and has similar height, serving score and endurance, however her final rating is better because slower times add more points. Since we are talking about volleyball teams, we probably want to reward quicker sprinting times.

In some cases, it may not be the highest or lowest that becomes ideal but rather a mid-range value. For example, if we are looking at ideal holiday destinations and want to take the climate into account, the best temperature might be described as one that is "not too hot and not too cold".

2.1.2   Consistent Scales

Megumi has the lowest score here—but that's largely due to her height. If we look at the range of the heights, not only are the values higher, but Megumi and Kayoko differ by 43 cm. That means in the final arithmetic mean, Megumi will already be almost 11 points lower than Kayoko.

Recall that we usually are able to interpret the arithmetic mean in the same units as the inputs, however here obviously there is no scale over which the arithmetic mean makes sense other than just a score or 'points'.

2.1.3   Differences in Distribution

Further to the problem of scale, sometimes the distribution should also be taken into account (Fig. 2.1).

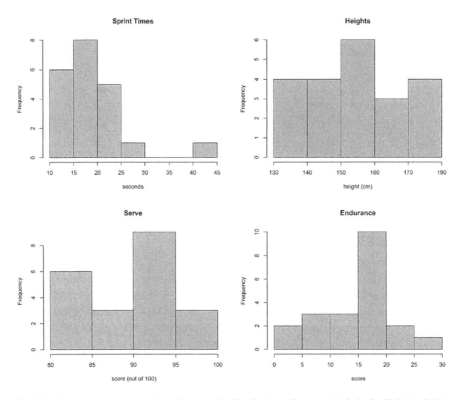

Fig. 2.1 Histograms generated in R showing the distribution of each variable in the Volleyball data

With sprinting times, the majority of students performed their sprint in between 10 and 20 s, however we would think the difference between a 10 and 15 s runner is more significant than the difference between a 20 and 25 s runner. On the other hand, the student who ran the sprint in 40 s may be much slower than everyone else, but we might not want this score in just one variable to have an undue influence on the final result.

Correcting for distributions can be very difficult. As we will see, there are some standard approaches we can use, however none of these is fool-proof and so we are often required to make a judgement call about what is reasonable.

In general, we should consider the following:

- If we are using *aggregation functions*, we need to bear in mind the interpretation of monotonicity. We need to ensure that it makes sense that an increase in the input should result in an increase to the output (e.g. for the students playing volleyball we need to transform the sprinting times);
- Scale and distribution of the data may differ. We may want/need to have all the data transformed to a particular interval, and we also might want to ensure that increases are treated similarly no matter where they are on that scale;

- Specific to averaging aggregation functions, we might want to consider the role that idempotency of the function plays. Does it make sense that a score of 0.8 for every input should result in an output that is also 0.8? For example, in grading students on entrance exams to university, a student that scores above average in every subject might be very rare, so we might actually want a score of 80 % on every subject to correspond with a score of about 95 % overall.
- The data might not be numeric at all. If we assign numeric values, are these reasonable and justified?

2.2 Background Concepts

In this chapter we will start to refer to more than just vectors of values to denote a set of inputs. We will make use of data organized in tables/arrays or matrices.

2.2.1 Arrays and Matrices (X)

In the students' volleyball data, each student (or observation/instance) could be considered to have a vector of attributes. As an example, for the data relating to Yukiko (the 6th student), we have $\mathbf{x} = \langle 19.17, 158, 83, 12 \rangle$. However in some cases we will also want to refer to the data relating to each variable, for example, the *Endurance* variable could be associated with the vector of length 20, $\mathbf{x} = \langle 17, 20, 17, 16, \ldots, 26, 16 \rangle$.

As well as Yukiko's height, 158, being an entry of either Yukiko's row vector or the *height* column vector, it can also be thought of as an entry of the table. We will refer to data made up of rows or columns of multiple vectors as 'matrices' or 'arrays'. We will usually use capital bold letters, e.g. $\mathbf{X}$, to refer to them and specify their dimensions $m \times n$, where m is the number of rows and n is the number of columns. For example, excluding the variable recording the students' names, we can refer to the volleyball data as a 20×4 matrix and label it $\mathbf{V}$.

2.2.2 Matrix/Array Entries (x_{ij})

We can then identify each entry in terms of its row and column. We write x_{ij} (or $x_{i,j}$ if there are more than 10 columns/rows to avoid ambiguity) to refer to the entry in the i-th row and j-th column. The entry $v_{3,4}$ in the volleyball data then is the 3rd value in the 4th column or the 4th value in the 3rd row, i.e. it corresponds with Megumi's endurance score.

2.2.3   Matrix/Array Rows and Columns ($\mathbf{x}_i$, $\mathbf{x}_j$)

However we can also refer to entire columns or rows. We will use the notation $\mathbf{x}_i$ to refer to row vectors and $\mathbf{x}_j$ to refer to columns. Sometimes it wouldn't be clear whether an instance $\mathbf{x}_3$ refers to a row or a column, so in cases where it could be ambiguous, we will specify instances as explicitly relating to i or j, e.g. $\mathbf{v}_{i=4}$ is the data relating to Sakura while $\mathbf{v}_{j=3}$ relates to the vector of data pertaining to the *Serving* variable.

2.3   Negations and Utility Transformations

Negation functions transform the data so that high values become low and low values become high. If the values are given over the unit interval (between 0 and 1), then the **standard negation** is given by

$$N(t) = 1 - t.$$

The standard negation is an example of a strict negation (e.g. see [2, 3, 6]), the term 'strict' meaning the same thing it does when we talk about monotone increasing behaviour.

Definition (informal) 2.1 (Strict Negation). A strict negation is a strictly decreasing function of one variable that has a maximum and minimum output that are the same as the domain of the inputs. So if the data we are transforming ranges from 1 to 10, then after we apply the transformation, the data should still range from 1 to 10, however the low values will now be high and the high values low.

Definition 2.1 (Strict Negation). A strict negation N defined over a real interval $[a, b]$ is a function that:

- is monotone decreasing, i.e. if $x < y$ then $N(x) > N(y)$; and
- satisfies boundary conditions $N(a) = b$ and $N(b) = a$.

Side Note 2.1 *In research literature, there is also the concept of a* strong negation, *which is one that satisfies the property of "involution". This means that if we perform a negation of the negation then we get the original value.*

$$N(N(t)) = t.$$

> **Notation Note** Function of a function
> When we have a function of a function, written either as something like $f(g(x))$ or $f \circ g(x)$, it means that we replace the variable in f with the whole function $g(x)$. For example if $f(x) = x^3$ and $g(x) = x^2 + 1$ then $f(g(x)) = (g(x))^3 = (x^2 + 1)^3$.
> In the case of the standard negation, $N(t) = 1 - t$, and so taking the standard negation *of* the standard negation will give $N(N(t)) = 1 - (1 - t) = 1 - 1 + t = t$.

However as is the case with our volleyball team data, we will often be dealing with data that is not expressed over the unit interval, and we therefore need to pay attention to the range of values. For the girls' volleyball team, we want to transform the *Sprint* variable. The data ranges from 10.96 to 40.29. It would be fine to use a negation like:

$$N(t) = 40.29 - t + 10.96 \quad \text{or} \quad N(t) = 51.24 - t$$

The result of this negation is depicted in Fig. 2.2.
We can note that this satisfies the boundary conditions since $N(10.96) = 40.29$ and $N(40.29) = 10.96$.
In general, if our interval is $[a, b]$, we can use

$$N(t) = b - t + a.$$

Negation transformation of Sprint data

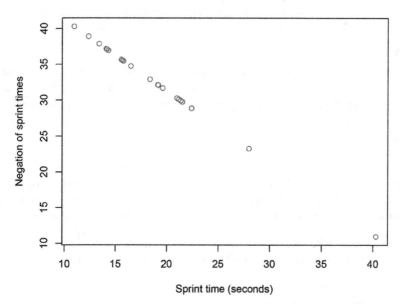

Fig. 2.2 Scatterplot showing sprint times and their negation which can then be interpreted so that 'higher means better'

Example 2.1. Show that the negation $N(t) = 51.24 - t$ over the interval $[10.96, 40.29]$ is a *strong negation*.

Solution. Since the function uses $-t$, it is clear that it will be monotone decreasing since the larger the input, the smaller the output. We just need to verify that it satisfies the property of involution.

Let's check:

$$N(N(t)) = 51.24 - N(t)$$
$$= 51.24 - (51.24 - t)$$
$$= 51.24 - 51.24 + t$$
$$= t$$

So this is a strong negation over the given interval.

We can have other negations too, for example if our data is over the unit interval, $N(t) = 1 - t^2$ is a strict negation (but not a strong negation) and $N(t) = \sqrt{1 - t^2}$ is a negation that is both strong and strict.

Side Note 2.2 *With our sprint data, there is no particular reason why we need the transformed data to be given over the same range. We would now interpret the times as 'time under 51.24' or, more generally, that the values now represent a kind of utility or points/rewards system. In the following subsection we will look at getting all of our data to a consistent scale—so sometimes we may first transform the data to a unit interval and then apply a negation (or vice versa).*

As well as our data going in the wrong direction, we might have situations where 'good' might actually refer to some intermediate value, with values being worse as they get further away. In this case, we are looking for the "just right" value (i.e. as with the porridge and beds in the *Goldilocks and the Three Bears* children's story[2]) (Fig. 2.3).

Whether we use the standard negation or something more complicated, our main concern is usually with the tendency of our output to increase or decrease with respect to our input variable. The standard negation will preserve most of the characteristics of our data distribution, e.g. if the majority of data lie within a given range then this density should stay the same if we only use a standard negation.

Example 2.2. What will be the transformed values for the *Sprint* variable for the first 8 students in the volleyball data using $N(t) = 51.24 - t$?

---

[2]On finding the three bears' uneaten bowls of porridge, Goldilocks tastes and remarks that the father's bowl of porridge is too hot, the mother bear's is too cold, but the baby bear's bowl is "just right" (so she eats the whole bowl).

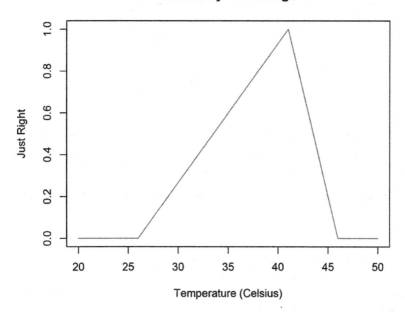

Fig. 2.3 Plot showing a transformation from raw temperatures to a degree of suitability, where porridge served too hot (above 46°) or too cold (below 26°) has a zero level of 'suitability'. The transformed scores would then satisfy the 'higher is better' property needed for monotonicity to make sense

Solution. Applying the transformation leads to the transformed scores as follows,

Student	Sprint (time under 51.24 s)	Student	Sprint (time under 51.24 s)
Mizuho	35.46	Izumi	40.28
Yukie	30.09	Yukiko	32.07
Megumi	36.94	Yumiko	32.89
Sakura	31.65	Kayoko	37.15

Example 2.3. Suppose we are evaluating employees in terms of customer feedback, their average service time and the number of customers they have handled. We have 5 employees to score with the following data. The customer feedback has been recorded as either positive or negative.

	Customers	Positive	Negative	Avg. time
Pete	12	1	6	16.2
John	73	10	6	18.7
Richard	59	38	12	7.8
George	62	10	6	6.8
Paul	44	35	7	7.3

What are appropriate negations that could be used for this data?

Solution. We can assume that since the number of customers handled and the number of positive feedback responses should contribute positively to an overall evaluation, we don't need to apply negations for these.

For negative feedback responses, the data ranges from 6 to 12. We can use

$$f(t) = 18 - t$$

so that the values in the transformed variable increase with fewer negative responses. The new data will be $\langle 12, 12, 6, 12, 11 \rangle$.

For the average time, the data ranges from 6.8 to 18.7. Since 'quicker' is usually better, we can transform this data using

$$f(t) = 25.5 - t.$$

The transformed values will be $\langle 9.3, 6.8, 17.7, 18.7, 18.2 \rangle$.

Note that our transformed data maintains the same data range and we have not yet addressed the different scale of the variables.

2.4   Scaling, Standardization and Normalization

If all our data can take values over a consistent range and have a consistent interpretation, for example, when we are finding the average measurement for a group with respect to one variable (e.g. height, long jump, amount of time spent sleeping etc.) then there would usually be no need to change the scale. We can take an average (using the arithmetic mean, the geometric mean, the median, or a number of other averages) and the output can be interpreted in the same units.

On the other hand, if the source and type of inputs vary (as is the case with each vector associated with a student in the volleyball data), then this is no longer possible. If we do not have a consistent scale, more varied inputs may have an undue influence in the aggregation step. We should also bear in mind that, whether or not the scale is consistent, if the type of inputs differs we should be careful about how we interpret our aggregated value. Usually it comes to represent a 'score' or overall evaluation, in which case our inputs should then be interpreted as contributing partial scores or evaluations with respect to different criteria.

The simplest technique for transforming each variable, which also preserves the distribution features, is to use linear transformations that scale the data to the unit interval. We will refer to this process as linear feature scaling.

Definition (informal) 2.2 (Linear Feature Scaling). When we have different variables or features, linear feature scaling for each variable is the process of transforming the data so that it ranges over the unit interval using only addition/subtraction and multiplication. A value a is subtracted from each entry and then we multiply by a factor b. For example, if we had heights ranging from 150 to 200 cm and weights ranging from 50 to 60 kg:

For the heights—we could subtract 150 (our a value) so that now the values range from 0 to 50, and then divide by 50 (our b value). Now all the values will be scaled to range from 0 to 1.

For the weights—we could subtract 50 (our a for this feature) and then divide by 10 (our b). This would mean both variables/features now have a consistent scale of 0 to 1.

Definition 2.2 (Linear Feature Scaling). For a set of values $\mathbf{x}_j = \langle x_{1,j}, x_{2,j}, \ldots, x_{m,j} \rangle$ relating to a single feature, we let $a = \min(\mathbf{x}_j)$ and $b = \max(\mathbf{x}_j) - \min(\mathbf{x}_j)$. Each $x_{i,j}$ in $\mathbf{x}_j$ can be scaled so that it takes a new value $x'_{i,j}$ over the unit interval using the transformation $x'_{i,j} = f(x_{i,j})$, where f is the single-variate function

$$f(t) = \frac{t - a}{b}.$$

Notation Note Transformations
With the index notation here, the j remains the same. This indicates that we are only considering the data pertaining to one feature (one of the columns in our matrix/array). The values for a and b would generally differ for each j, but they should be the same for each i when j is fixed (assuming that columns represent features and rows represent cases). We have also used the notation x' to denote a transformed or 'updated' value. We could also write something like,

$$x_{\text{new}} = \frac{x_{\text{old}} - x_{\text{min}}}{x_{\text{max}} - x_{\text{min}}}.$$

We still use t as the variable in the function $f(t)$. Just remember that the entry of the matrix $x_{i,j}$ is what we substitute for t when we write $f(x_{i,j})$.

Example 2.4. What will be the transformed values for the first 8 students if we scale the *Height* variable to the unit interval using linear feature scaling?

Solution. The tallest height (for the 20 students) is 177 and the shortest is 134. We therefore can use $f(t) = \frac{t-134}{43}$ to give the following heights.

Student	Height (score between 0 and 1)	Student	Height (score between 0 and 1)
Mizuho	0.33	Izumi	0.26
Yukie	0.30	Yukiko	0.56
Megumi	0	Yumiko	0.53
Sakura	0.93	Kayoko	1

As mentioned already, linear feature scaling will preserve the essential features of the distribution, however sometimes this can cause problems. For example, if we have a single very large outlier, then now most of our values will be scaled so that they are close to zero. If you have studied statistics, you might have come across the idea of standardization or Z-scores.

Definition (informal) 2.3 (Standardization). Standardization usually refers to the process of subtracting the mean and dividing by the standard deviation for each value, which centers the data at zero and scales the standard deviation to 1. Applying this technique to different variables makes them commensurable, and we then essentially interpret them in terms of how abnormal they are.
*Having a good feel for interpreting standardized values and averages of standardized values requires familiarity with the normal distribution, which is beyond the scope of the topics we cover but can be found in any introductory statistics book (e.g. see [4] for more information as well as a number of standard statistical techniques for transforming data).

Definition 2.3 (Standardization). For an input vector $\mathbf{x}_j = \langle x_{1,j}, x_{2,j}, \ldots, x_{m,j} \rangle$ where $SD(\mathbf{x}_j) = \sqrt{\sum_{i=1}^{n} \frac{(x_{i,j}-\mu)^2}{n-1}}$ is the sample standard deviation of $\mathbf{x}_j$ (or the true mean (μ) and standard deviation (σ) may be known *a priori* for the wider population of data observations), standardization involves transforming each $x_{i,j}$ using $x'_{i,j} = f(x_{i,j})$, where

$$f(t) = \frac{t - AM(\mathbf{x}_j)}{SD(\mathbf{x}_j)}.$$

Notation Note Transformations
Once again, we could express this transformation as an update process from old to new values,

$$x_{\text{new}} = \frac{x_{\text{old}} - AM(\mathbf{x}_j)}{SD(\mathbf{x}_j)}.$$

This is an appropriate transformation if our data all follow a normal distribution and differ only in terms of their mean and standard deviation. Following the

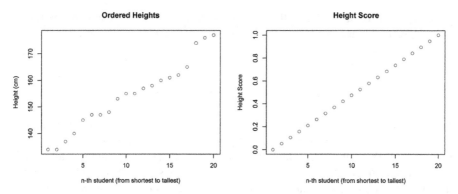

Fig. 2.4 The student data for the height variable before (*left*) and after (*right*) rank-scaling

standardization step, our data would then usually fall between -2 and 2, with approximately 5 % lying above or below. If we wanted our data to lie between 0 and 1 we could then use linear feature scaling on the standardized data. Assuming there are no extreme outliers, one transformation that in most cases will allow normally distributed data to be scaled to the unit interval is:

$$f(t) = 0.15 \left(\frac{t - \mathrm{AM}(\mathbf{x})}{\mathrm{SD}(\mathbf{x})} \right) + 0.5$$

The 0.15 multiplying the usual standardization formula will first scale the data so that 99.7 % of the data should fall between -0.45 and 0.45 (three standard deviations) and then adding 0.5 will shift the interval to $[0.05, 0.95]$.

You will note the similarity between the form of these two types of transformation, which can both be considered as kinds of 'normalization'. They only involve simple operations on the data and essentially preserve the features of the distribution in a relative way. Another simple transformation that can be applied is rank-scaling (Fig. 2.4). In this case, we are only interested in the relative order of the data—not how far apart they are. Rank-scaled data also enables the data to be easily interpreted in terms of percentiles, i.e. a score of 0.9 means the value is higher or 'better' than 90 % of the other evaluations.

> **Definition (informal) 2.4 (Rank-Scaling).** Rather than the actual scores, the relative ordering or ranking might be what is more important. In rank-scaling we first order the data and then allocate scores based on the percentage of data below each value. E.g. for $\mathbf{x} = \langle 102, 37, 10, 39, 28 \rangle$ we would have $\langle 1, 0.5, 0, 0.75, 0.25 \rangle$.

> **Definition 2.4 (Rank-Scaling).** For an input vector $\mathbf{x}_j = \langle x_{1,j}, x_{2,j}, \ldots, x_{m,j} \rangle$, let $O_j(x_{i,j})$ denote the rank of $x_{i,j}$ with respect to the other entries in $\mathbf{x}_j$, so that

$O_j(x_{i,j}) = 1$ means that $x_{i,j}$ is the 'best' or highest score, $O_j(x_{i,j}) = 2$ means $x_{i,j}$ is the second highest, and so on. We can transform each $x_{i,j}$ into a score out of 1 using $x'_{i,j} = f(x_{i,j})$, where

$$f(t) = \frac{m - O_j(t)}{m - 1}.$$

We can also use the ranks themselves if we prefer. Ranking transformations can be useful for data that is ordinal or naturally refers to a scale that is not necessarily numerical. If there are ties, usually each alternative in the tied position is allocated the average of their ranks. For example, if 3 alternatives 4th, 5th and 6th were tied, they would each receive a rank of 5. If two alternatives in positions 8 and 9 were tied, they would each receive a rank of 8.5.

> **Side Note 2.3** *We should bear in mind, however that we are imposing a numerical interpretation that may not be valid. For example, in iMDb movie ratings, it is not necessarily the case that the difference between 9/10 and 10/10 is the same as the difference between 6/10 and 7/10. What we need to be aware of when we perform rank-transformations (or in fact any kind of utility transformation) is that we are essentially awarding a 'score' or a 'partial score' based on that variable or feature. Whenever we aggregate data with averaging functions, low values in one variable can be compensated for by high values in others. We should think carefully about the implications of our transformations with respect to this property.*

2.5 Log and Polynomial Transformations

Other common transformations used as part of some statistical techniques includes the use of increasing functions like

$$f(t) = \ln t,$$

or

$$f(t) = t^2.$$

Such functions can help if data are exponentially distributed or skewed.[3]

The log function in particular is useful for data that can have a few very large inputs (Fig. 2.5). Incomes, populations in ecology and academic journal citations are all examples of data that could exhibit such properties.

---

[3]Characterizing distributions is beyond what we cover, so we will just quickly note that exponentially distributed data has most of the data gathered towards very low values with fewer and fewer high values. The high values, however can be *very* high.

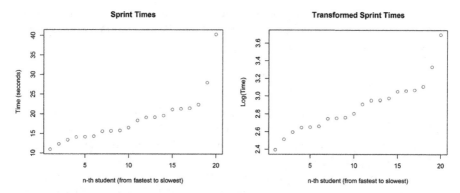

Fig. 2.5 A log transformation of the sprint variable. The relative difference between lower values is increased and higher values is decreased. The natural log (with the base e) is used

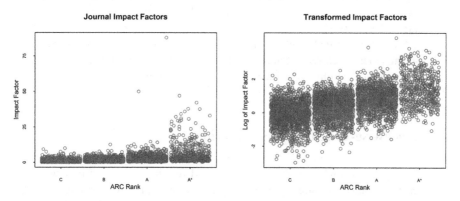

Fig. 2.6 Journal impact factor data for 5311 journals organized by the 2011 ARC rankings (and randomly within ranks). The log transformation (*on the right*) makes it easier to see the spread of the different classes

Although the general difference between the values is somewhat maintained, values toward the lower end of the scale become more spread out while higher values are pushed closer together. Another example is shown in Fig. 2.6. In this case, the data relates to the Australian Research Council journal rankings released in 2011 and the impact factors (the ratio of citations to a journal's papers within the last two years to articles published) are shown with journals organized into their evaluated rankings.[4] The transformation allows for the outliers to become less pronounced.

Polynomial function transformations $f(t) = t^2$, $f(t) = t^{\frac{1}{2}}$ can have a similar (but less drastic) effect, allowing skewed distributions to become more symmetric.

---

[4]See [1] for a study on predicting these rankings from citation indices. The data shown is available at http://aggregationfunctions.wordpress.com/data-sets.

For t^p, $p > 1$ can be used when there are fewer very high values (positive skew), while $0 < p < 1$ can be used if the majority of data is gathered in the high range with fewer very low values (negative skew).

Example 2.5. For which of the variables in the volleyball data would it make sense to use a log transformation?

Solution. Knowing when to use a log transformation is not straightforward. In some statistical approaches when looking for relationships between variables, the aim is usually to use transformation functions like log and then check if the relationship appears linear.

For our sprint times (before being transformed by a negation) we can see that many of the values are clustered together toward the lower end of the scale and there are is one isolated value in the higher end. A log transformation will spread out the lower values and bring the outlier closer to the main group, however it might be preferable to just remove the outlier—without it, the data is more or less normal.

In general, log transformations make sense when the data looks similar to the sprint times when represented as a histogram, however with the majority of values in the first interval close to zero (or the minimum).

2.6 Piecewise-Linear Transformations

If we have good knowledge about the data and distribution, we might want to apply custom transformations that spread out the data in a way tailored to our application. The easiest of these (although it can still be complicated) is to construct a transformation from either linear or even quadratic 'pieces'. We already saw a similar tactic used for transformation of the porridge temperatures to a degree of how "Just Right" they are.

When doing this, we usually split our domain into subdomains and our aim is that on the border of each domain the transformation functions are equal.

Definition (informal) 2.5 (Piecewise-Linear Transformations). The piece wise-linear transformations we use will usually be monotone functions where the domain is split into intervals and a different linear function (i.e. $f(t) = mt + c$) is used to transform the data over each interval. The functions should connect on the border of the domains for the transformations to be continuous. For example, if we split the unit interval in half, and use $f(t) = 0.4t$ over the sub-interval $[0, 0.5)$, and $f(t) = 1.6t - 0.6$ over the sub-interval $[0.5, 1]$, then the effect will be that the lower half of the data is pushed closer together and the upper half will be stretched.

The following allows us to scale the data to the unit interval with two linear functions (see Fig. 2.7).

Fig. 2.7 Diagram of a piecewise-linear function with 2 pieces. Everything between a and b is linearly scaled to range between 0 to q and everything from b to c scales to a value between q and 1

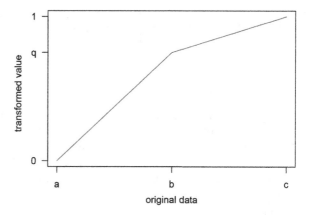

Definition 2.5 (Piecewise-Linear Transformations (2 Pieces)). For a set of evaluations $\mathbf{x}_j = \langle x_{1,j}, x_{2,j}, \ldots, x_{m,j} \rangle$ given over $[a, c]$, we split the domain into two sub-intervals, $[a, b)$ and $[b, c]$. Let $q \in [0, 1]$ be the transformed value we want our variable to take when $x_{i,j} = b$. Letting $x'_{i,j} = f(x_{i,j})$ with the following piecewise function scales the data to the unit interval.

$$
f(t) = \begin{cases}
q\frac{t-a}{b-a} & , \quad a \le t < b, \\[2ex]
q + (1-q)\frac{t-b}{c-b}, & b \le t \le c.
\end{cases}
$$

Notation Note Cases

The large '{' indicates cases, while the right hand inequations $a \le t < b$ and $b \le t \le c$ specify intervals that tell us when each case should be used. When our input is less than b, we use the first function and the second when it is b or above. The round bracket in the interval notation is the same as using a strictly less than symbol '<', i.e. indicating that the value is not included in the interval but everything below it is. We need to have one of the borders open (with a round bracket) to make it clear which function we should use when $x_{i,j} = b$, however we could also use $[a, b]$ and $(b, c]$. Since both functions should be equal at b, it doesn't matter which.

Similarly we can create a piecewise-linear transformation if want to split our function into three sections.

Definition 2.6 (Piecewise-Linear Transformations (3 Pieces)). For a set of evaluations $\mathbf{x}_j = \langle x_{1,j}, x_{2,j}, \ldots, x_{m,j} \rangle$ given over $[a, d]$, we split the domain into three sub-intervals, $[a, b)$ and $[b, c)$ and $[c, d]$. Let $q \in [0, 1]$ be the transformed value we want our variable to take when $x_{i,j} = b$ and $r \in [q, 1]$ be the value we want our variable to take when $x_{i,j} = c$ with $a < b < c < d$. Then we can apply the following transformation function.

$$f(t) = \begin{cases} q\frac{t-a}{b-a} & , \quad a \le t < b, \\ q + (r-q)\frac{t-b}{c-b}, & b \le t < c, \\ r + (1-r)\frac{t-c}{d-c}, & c \le t \le d. \end{cases}$$

We can continue this pattern for any number of sub-intervals. We just need to have a set of points where we know what the transformed output should be.

Whenever the multiplier is less than 1 over a given sub-interval, the data will be pushed together, while whenever the multiplier is greater than 1, the values will be spread further apart from one another. You will notice that the fractions involved in the expressions are of a similar form to the linear feature scaling and standardization formulas.

The following piecewise function was used with journals data in [1] to perform a custom scaling of the impact factors, which were later used along with other variables to estimate the overall ranking of each journal. In this ranking exercise, the A* ranked journals were considered to be in the top 5 % of journals, the A ranked journals were in the top 20 %, B in the top 50 % and C otherwise. The domain was split into sub-intervals corresponding with the median impact factor score for each journal and the values in the unit interval these values were chosen so that the percentiles values would fall between them, i.e. $y_C = 0.3, y_B = 0.7, y_A = 0.9$.

$$x'_{i,j} = \begin{cases} y_C\frac{x_j - \min(\mathbf{x}_j)}{C_{Med} - \min(\mathbf{x}_j)}, & x_j < C_{Med}; \\ y_C + (y_B - y_C)\frac{x_{i,j} - C_{Med}}{B_{Med} - C_{Med}}, & x_j < B_{Med}; \\ y_B + (y_A - y_B)\frac{x_{i,j} - B_{Med}}{A_{Med} - B_{Med}}, & x_j < A_{Med}; \\ y_A + (1 - y_A)\frac{x_{i,j} - A_{Med}}{A*_{Med} - A_{Med}}, & x_j < A*_{Med}; \\ 1 & \text{otherwise.} \end{cases} \qquad (2.1)$$

The plot of one of the piecewise functions created in this way is shown in Fig. 2.8. Figure 2.9 shows the impact factors and transformed scores separately so that differences in the distribution can be more clearly seen.

Example 2.6. Suppose we have data for one variable that ranges from 82 to 109, how can we transform it with a piecewise-linear function such that anything above 100 has a value of between 0.8 and 1, while anything below 100 ranges from 0 to 0.8?

Solution. We can use two functions, one to operate over the domain [82, 100] and one for values the values in (100, 109]. Using the formula for a 2-piece function where $[a, b] = [82, 109]$ and $q = 100$ we will have,

$$f(t) = \begin{cases} 0.8\frac{t-82}{18}, & t \le 100, \\ 0.8 + 0.2\frac{t-100}{9}, & \text{otherwise.} \end{cases}$$

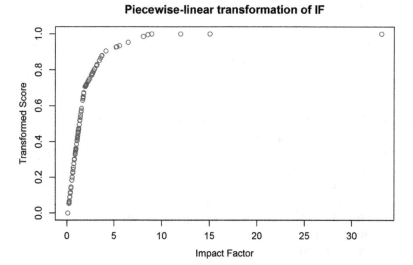

Fig. 2.8 Plot showing an example of a piecewise-linear transformation of the form in Eq. (2.1). For a sample of 100 journals, we assume the medians for each rank are denoted by 0.797, 1.877, 3.938 and 8.779 respectively, then assign these our desired outputs of 0.3, 0.7, 0.9 and 1

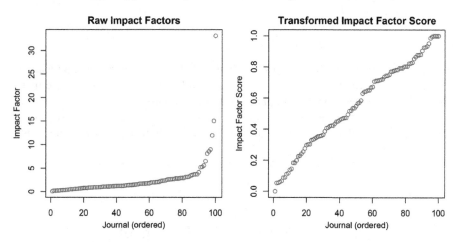

Fig. 2.9 Plots showing the ordered data for the raw impact factors (*left*) and transformed scores (*right*). The effect is that the relative differences for the lower 85 journals are increased while those of the upper 15 are pushed closer together

We can see that at $t = 100$, both values give the output of 0.8, while anything above will be between 0.8 and 1.

2.7 Functions Built from Transformations

Transformation functions can also be used in the construction of new functions, whose properties can be interpreted in light of the transformations.

One such function is called the **dual** aggregation function, which is built from a negation.

Definition (informal) 2.6 (Dual Aggregation Function). The dual aggregation function uses a negation (usually the standard negation). First, all of the inputs are transformed using that negation, the data is aggregated, and then finally the negation is used on the result. For example, using the arithmetic mean and the standard negation on the inputs $\langle 0.3, 0.9 \rangle$, the negation gives $\langle 0.7, 0.1 \rangle$, the average of these is 0.4, and using the negation again gives 0.6. Note that this is the same as the average of the two original values (this always happens for the arithmetic mean but not in general for duals of other aggregation functions).

Definition 2.7 (Dual Aggregation Function). For an aggregation function A with inputs given over the unit interval $[0, 1]$, its dual A^d (using the standard negation) is given by:

$$A^d(\mathbf{x}) = 1 - A(1 - x_1, 1 - x_2, \ldots, 1 - x_n).$$

For averaging aggregation functions, the dual will also be an averaging aggregation function, so we still have $A(a, a, \ldots, a) = a$ and $A(b, b, \ldots, b) = b$ and the final output bound between the minimum and maximum inputs.

In fact the dual of the arithmetic mean is the arithmetic mean itself, and so in this case, taking the dual does nothing at all. We can demonstrate this for two arguments.

$$AM^d(x_1, x_2) = 1 - AM(1 - x_1, 1 - x_2)$$

$$= 1 - \frac{1}{2}(1 - x_1 + 1 - x_2)$$

$$= 1 - \frac{2}{2} + \frac{x_1 + x_2}{2} = \frac{1}{2}(x_1, x_2).$$

The arithmetic mean is not affected because it treats inputs the same whether they are at the higher or lower end of the input range. However if we have a function that tends to be affected more by high inputs, its dual will be more affected by low inputs instead. As an example, the dual of the geometric mean has an absorbent element of 1 (i.e. if *any* of the inputs are 1, then the output will immediately be 1) (Fig. 2.10).

The dual function essentially exhibits reciprocal properties to the original function, but it remains monotone increasing.

GM(x)   **Dual of GM(x)**

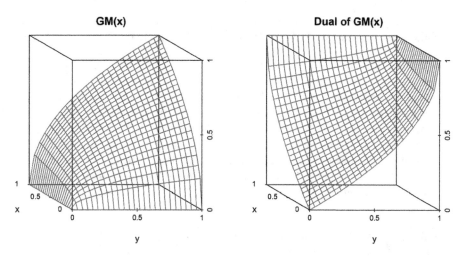

Fig. 2.10 The geometric mean (*left*) and its dual (*right*). The dual has an absorbent element of 1

Example 2.7. Define the dual of the geometric mean explicitly for two arguments.

Solution. Using the formula for construction of the dual with the geometric mean $GM(x_1, x_2) = \sqrt{x_1, x_2}$ gives,

$$GM^d(x_1, x_2) = 1 - \sqrt{(1 - x_1)(1 - x_2)}.$$

2.8   Power Means

Previously we saw that transforming data could be used to help address distributions that were skewed or included extreme values. There are special families of means built from similar transformations.

The power means constitute one such family, based on a parameterized transformation $f(t) = t^p$ where the parameter p can be any real number from negative infinity to infinity. Lower values of p result in outputs that are more sensitive to lower inputs, while high values of p result in power means that tend more toward the higher inputs. Note here that as opposed to our previous application of transformations, here all the inputs are transformed in the same way.

The difference between using transformations as we have done previously and using the arithmetic mean on these transformed values is that the power mean makes use of an inverse function, which returns outputs to their original scale. The upshot of this is that the resulting function will be idempotent and averaging, meaning that we can still interpret the output in the same way we interpret the inputs, and potentially use the same units if the inputs are all of the same type.

Definition (informal) 2.7 (The Power Mean). When calculating the power mean, we first transform all the inputs using $f_1(t) = t^p$, after which we take the arithmetic mean, and finally, use the transformation $f_2(t) = t^{1/p}$ on the result. This latter function f_2 is the inverse of f_1. As an example, if we have $\mathbf{x} = \langle 2, 3 \rangle$, then using the power $p = 2$ we would find the average of 2^2 and 3^2, which is $(4+9)/2 = (6.5)$. Raising 6.5 to the power of a half (or taking the square root) gives approximately 2.55, which is slightly higher than the arithmetic mean of 2 and 3 (which would be 2.5). Here we assume that all of the inputs are greater than or equal to zero.

Definition 2.8 (The Power Mean). For an input vector $\mathbf{x} = \langle x_1, x_2, \ldots, x_n \rangle$, with $x_i \geq 0$ for all i, the power mean is

$$PM_p(\mathbf{x}) = \left(\frac{1}{n} \sum_{i=1}^{n} x_i^p \right)^{\frac{1}{p}} = \left(\frac{x_1^p + x_2^p + \cdots + x_n^p}{n} \right)^{\frac{1}{p}} .$$

Notation Note Powers
Remember that raising a value to a fractional power is the same as taking the n-th root. Note here as well that we are not thinking about the power mean transformations x^p in terms of scaling or transforming variables, since these inputs may correspond with different criteria. In general, we assume that any scaling or normalization has already taken place to make the data commensurable.

The power mean is not just one function but a family of functions, with each member being determined by the value of p.

For $p = 1$, we obtain the arithmetic mean; for $p = -1$, we have the harmonic mean; and for $p = 0$ (a limiting case[5]) we actually obtain the geometric mean. Furthermore, as $p \to -\infty$ we approach the minimum function and $p \to \infty$ we approach the maximum. We summarize these in the table below.

---

[5]By 'limiting case' we are making reference to the mathematic concept of the limit. In the case of $p = 0$, as the parameter p gets closer and closer to 0, the outputs of the power mean become more and more similar to the outputs of the geometric mean, however we could never have p being equal to exactly 0, because then we would have $\frac{1}{0}$ as our fractional power and dividing by zero is not allowed. A value of $p = 0.00000001$ will give results that are numerically very similar to using the standard formula for the geometric mean. We often write "as $p \to 0$", which means "as p approaches zero".

Special case	p	
Maximum	∞	$\max(x_1, x_2, \ldots, x_n)$
Quadratic mean	2	$\sqrt{\dfrac{1}{n}\displaystyle\sum_{i=1}^{n} x_i^2}$
Arithmetic mean	1	$\dfrac{1}{n}\displaystyle\sum_{i=1}^{n} x_i$
Geometric mean	0	$\left(\displaystyle\prod_{i=1}^{n} x_i\right)^{1/n}$
Harmonic mean	-1	$\left(n\displaystyle\sum_{i=1}^{n} x^{-1}\right)^{-1}$
Minimum	$-\infty$	$\min(x_1, x_2, \ldots, x_n)$

Side Note 2.4 *The power mean is sometimes referred to as a generalized mean, since it includes many means as special cases.*

Power means are averaging aggregation functions for any value of p. This means they are always monotone and satisfy the boundary conditions. They are also homogeneous for any p, however it is only in the case of the arithmetic mean ($p = 1$) that they will be translation invariant.

Guided by our special cases, we can consider how the function will treat inputs depending on the value of p.

- Larger values of p (and greater than 1) will mean that the output of the function is more influenced by larger values in the input set than by smaller ones. So the output will be 'dragged' towards higher inputs. When p becomes infinitely large, the output will simply be the highest input value.
- Values of p below 1 will drag the function towards lower inputs. Furthermore, if $p \leq 0$, we obtain functions that will have an output of zero if any of the inputs are equal to zero (e.g. we already saw that the geometric mean ($p = 0$) and harmonic mean ($p = -1$) had an *absorbent element* of 0). Values of p between 0 and 1 will still tend toward lower values but won't have this absorbing element property.

Example 2.8. Compare the values of the power mean for $p = -5, 0, 1$ and 3 for the input vector $\mathbf{x} = \langle 0.3, 0.9, 1 \rangle$.

Solution. Calculating these values respectively gives outputs of approximately $0.373, 0.646, 0.733$ and 0.837. It is clear that while $p = -5$ pushes the output closer to the 0.3, as it is increased the output draws closer to the 1.

2.9 Quasi-Arithmetic Means

An even more general family of averaging functions is the quasi-arithmetic means. Rather than a transformation of the variable using $f(t) = t^p$, the transformation f can be of any form. It is then referred to as a generating function, which we will denote by $g(t)$. The generating function g can be almost anything that is defined over the domain of the inputs, provided it is monotone and has an inverse (e.g. with $f(t) = t^p$, the inverse was $t^{1/p}$, i.e. it is a function that 'undoes' the other). We can even use piecewise-linear functions. We will not focus in great detail on quasi-arithmetic means since the power means are broad enough for most of our needs, however the following definition is provided for those interested and further information can be found in any of the key references [2, 3, 5, 6, 9].

Definition 2.9 (The Quasi-Arithmetic Mean). For an input vector **x** and suitable generating function g, the quasi-arithmetic mean is given by

$$QAM_g(\mathbf{x}) = g^{-1}\left(\frac{1}{n}\sum_{i=1}^{n} g(x_i)\right).$$

Of course if g is t^p then we obtain the power means as a special case. As well as being a special case of the power mean, the geometric mean can also be expressed as a quasi-arithmetic mean with respect to the generator $g(t) = \ln t$. Rather than relying on the idea of the limit, in this case we can show that using $\ln t$ and its inverse recovers our original expression for the geometric mean as the product of all arguments raised to the power $\frac{1}{n}$. The key to this is the log law that adding logarithms together is the same is taking the log of their product, i.e. $\log a + \log b = \log(ab)$. If $g(t) = \ln t$, then $g^{-1}(t) = e^t$ and we will have:

$$GM(\mathbf{x}) = e^{\left(\frac{1}{n}\sum_{i=1}^{n}\ln x_i\right)}$$

$$= e^{\frac{1}{n}\ln\left(\prod_{i=1}^{n} x_i\right)} = e^{\ln\left(\prod_{i=1}^{n} x_i\right)^{\frac{1}{n}}}$$

$$= \left(\prod_{i=1}^{n} x_i\right)^{\frac{1}{n}}.$$

Notation Note Logarithms
The notation $\ln t$ means $\log_e t$ or log to the base e. We could actually use a logarithm with any base, since its inverse would still cancel out the way it does with $\ln t$ and e^t.

2.10 Summary of Formulas

Strict Negation

Over a real interval $[a, b]$, a strict negation N is a function that satisfies

$$\text{If } x < y \text{ then } N(x) > N(y) \quad \text{(monotonicity)}$$

(2.2)

$$N(a) = b, N(b) = a, \quad \text{(boundary conditions)}$$

Linear Feature Scaling

$$f(t) = \frac{t - a}{b},$$

(2.3)

$$a \leq \min(\mathbf{x}), \qquad b \geq \max(\mathbf{x}) - \min(\mathbf{x})$$

Standardization

$$f(t) = \frac{t - AM(\mathbf{x})}{SD(\mathbf{x})},$$

(2.4)

Rank-Scaling

$$f(t) = \frac{m - O(t)}{m - 1},$$

(2.5)

(m is the number of data, $O(t)$ is the ranking among the data relating to the variable we are transforming)

Piecewise-Linear Transformations (2 Pieces)

$$f(t) = \begin{cases} q\frac{t-a}{b-a} & , \quad a \leq t < b, \\ q + (1-q)\frac{t-b}{c-b}, & b \leq t \leq c. \end{cases}$$

(2.6)

Piecewise-Linear Transformations (3 Pieces)

$$f(t) = \begin{cases} q\frac{t-a}{b-a} & , \quad a \leq t < b, \\ q + (r-q)\frac{t-b}{c-b}, & b \leq t < c, \\ r + (1-r)\frac{t-c}{d-c}, & c \leq t \leq d. \end{cases}$$

(2.7)

Dual Aggregation Function

$$A^d(\mathbf{x}) = 1 - A(1 - x_1, 1 - x_2, \ldots, 1 - x_n) \tag{2.8}$$

The Power Mean

$$\text{PM}_p(\mathbf{x}) = \left(\frac{1}{n} \sum_{i=1}^{n} x_i^p \right)^{\frac{1}{p}} = \left(\frac{x_1^p + x_2^p + \cdots + x_n^p}{n} \right)^{\frac{1}{p}} \tag{2.9}$$

The Quasi-Arithmetic Mean

$$QAM_g(\mathbf{x}) = g^{-1} \left(\frac{1}{n} \sum_{i=1}^{n} g(x_i) \right) \tag{2.10}$$

2.11 Practice Questions

1. Consider the following wait and service times for 100 customers.

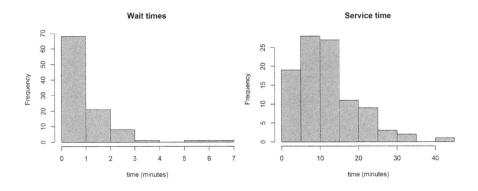

   What would be some potential transformations that could be used to scale these variables to the unit interval?
2. Below are histograms of the accuracy and computation time taken for varying classification parameters. We want to be able to aggregate the two scores in order to be able to determine the best overall classifier.
   Suggest some transformations that would be able to transform the variables so that it makes sense to take their average.
3. For the following piecewise function,

$$f(t) = \begin{cases} \frac{t}{5}, & t < 3, \\ \frac{3}{5} + 2\frac{t-3}{10}, & 3 \le t \le 5, \end{cases}$$

what will be the value when $t = 2$? How about when $t = 4$?

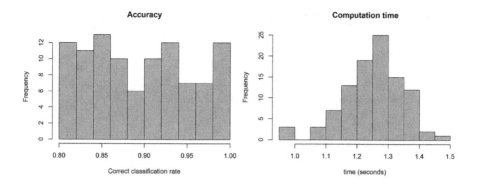

4. Is the power mean symmetric, homogeneous and translation invariant? Explain.
5. Does the power mean have absorbent elements? Explain.
6. Write out the power mean explicitly for 3 arguments when $p = 2$.
7. Write out the power mean explicitly for 2 arguments when $p = -4$.
8. If $\text{PM}_p(9, 10, 17, 16) = 13$ for some value of p, can we work out the value of $\text{PM}_p(12, 13, 20, 19)$ and $\text{PM}_p(18, 20, 34, 32)$ without knowing p?
9. If $p = -4$, can we determine the value of $\text{PM}_p(3, 0, 2, 8)$ without calculating?
10. Write out the explicit formula (i.e. without the $\sum$ symbol) for $\text{PM}_p(x_1 + 3, x_2 + 3)$.
11. If $\text{PM}_p(3, 7, 29, 45) = 7.162$, is the value of p likely to be greater than 1 or less than 1?
12. If $\text{PM}_p(3, 7, 29, 45) = 36.258$, is the value of p likely to be greater than 1 or less than 1?
13. If $\text{PM}_p(2, 3, 8, 3) = 7$, is the value of p likely to be high or low?

2.12   R Tutorial

Here we will start working with whole matrices of data and in R [8] learn how to implement transformations and power means.

2.12.1   Replacing Values

We learnt how to assign vectors and values in the previous chapter. If we want to change the value of just one entry in a vector, we can use the assign command and indicate the index in square brackets, e.g. to change the 5th entry in a vector a we would enter

```
a[5] <- 10
```

We can also replace multiple entries at once

```
a[c(5,7)] <- c(10,3)
a[3:5] <- c(1,0,1)
```

R Exercise 6 *Create the vector and change the entries.*

```
a <- rep(0,20)
a[5] <- 1
a[c(3,7,11)] <- c(2,6,1)
a[17:20] <- c(1,2,1,4)
```

Your expected output when you type a *and press enter should now be*

0 0 2 0 1 0 6 0 0 0 1 0 0 0 0 0 1 2 1 4

2.12.2 Arrays and Matrices

Sometimes we will need to consider whole datasets at a time. For this, rather than just vectors, we will need rows of vectors, or matrices and arrays.

We can build them step by step with the cbind() (column bind) and rbind() (row bind) functions.

```
a <- cbind(c(2,3,5,1,0), c(6,1,8,2,9))
b <- rbind(c(4,1), c(2,-2),c(5,6))
```

R Exercise 7 *Assign the vectors and perform the* cbind() *and* rbind() *operations.*

```
a <- c(1,2,3,7,9)
a <- cbind(a, c(21,2,1,5,6))
a <- cbind(a, c(2,-1,5,0,-1))
a <- cbind(a,c(1,9,7,2,1),array(6,5))

b <- c(3,6,1,92)
b <- rbind(b, c(3,2,1,8,9))
b <- rbind(b, c(4,1,12,1,2))
```

Note that for vectors like c(3,2,1,8,9), R doesn't treat it as either a row or a column when it's by itself and so it is flexible when it comes to combining vectors either as rows or as columns. However once we have a matrix, the rbind and cbind operations will come up with a warning if the row length is not the same—however it will still merge them (it just fills the remaining space by repeating the sequence of the row).

We can also create large $m \times n$ arrays and then input values later. We will use the array function for this.

> **Side Note 2.5** *Matrices can also be constructed using* `matrix()`, *however the array function will be the most straightforward to use at this time.*

As with the array function previously, the first entry is the default value that will populate the array, however this time we will use a vector `c(rows,columns)` to indicate how many rows and how many columns there needs to be. A matrix/array with 3 rows and 4 columns (prepopulated with 0s) would be

```
A <- array(0,c(3,4))
```

> **Side Note 2.6** *We also can have higher dimension arrays, however for the moment we will stick to two.*

Once we've created our array with default values, we can then proceed to fill it. We refer to the cells using `A[row,column]` so that the entry in the first row and the second column would be `A[1,2]`. We can also consider entire columns, e.g. the second using `A[,2]` and entire rows using `A[1,]`. Note that this corresponds with our x_i, x_j and $x_{i,j}$ notation. Using `A[,2]`, will treat the column as a normal vector, however we can also use `A[,2,drop=FALSE]`, which maintains its column structure.

R Exercise 8 *Create a 3×4 array and then assign values to different entries using the following.*

```
A[3,1] <- 4
A[1,] <- c(1,2,3,4)
A[,2] <- c(6,5,4)
A[3:4,2:3] <- array(-1,c(2,2))
```

Your final matrix should appear as

```
      [,1] [,2] [,3] [,4]
[1,]    1    6    3    4
[2,]    0    5   -1   -1
[3,]    4    4   -1   -1
```

2.12.3   Reading a Table

Often we will have data available from some other source, e.g. as an Excel spreadsheet. We can import this data using the `read.table()` function. The easiest way is to have the data in a txt or csv file. We need to know whether the entries are separated by commas or spaces and whether they have labels. In the simplest case, we can save the table to an array. If the file is in the same folder as our R working directory,[6] we can use the command

```
A <- read.table("thedata.csv")
```

If the data has headers (i.e. the first row is not entries but data labels), or is separated by commas, then we can add extra options.

```
A <- read.table("thedata.csv",header=TRUE,sep=",")
```

It is easiest if our data is already numerical data. In some cases, the data can be coded as text even if it is numbers. If columns in the data are numeric and R has trouble interpreting them as numbers (e.g. it says NA when you try to add 2, where NA indicates not available, i.e. a missing value), they can be extracted from the table using the as.numeric() command. The following exercise uses the write.table() command as well.

R Exercise 9 *Use the following to create and write a table to the working directory*

```
write.table(cbind(c("a",1,2),c(3,2,5)),"wordy.txt")
```

Now read the table from the file and assign it to my.table *using the* read.table() *command.*

```
my.table <- read.table("wordy.txt")
```

See what happens when you input the following.

Input	Expected output
my.table	V1 V2 1 a 3 2 1 2 3 2 5
my.table+2	V1 V2 1 NA 5 2 NA 4 3 NA 7
my.table[2,]+2	V1 V2 2 NA 4
as.numeric(my.table[2,])+2	3 4

---

[6]In RStudio, the working directory can be set using the Session→Set Working Directory option from the menu. In the standard R application on a Mac, the option to change the working directory is under Misc, while in Windows it is under File. You can also use the setwd() function directly from the console.

2.12.4 Transforming Variables

Simple transformations of variables can be achieved using operations that we have already learned. We can either choose to simply replace our values with the transformed ones or we can create a new matrix.

Let's first load our volleyball data[7] and then make a copy which we will store as "original".

```
V <- read.table("volley.txt")
original <- V
```

The second of these lines simply copies the table "V" and assigns the copy to `original`. The following command replaces the *Sprint* variable with transformed values according to our negation function.

```
V[,1] <- 51.24 - V[,1]
```

We've used `V[,1]` to access the whole first column and when we replace it with `51.24 - V[,1]`, each entry in the column is subtracted from 51.24. We can then transform it to the unit interval with the following

```
V[,1] <- (V[,1]-min(V[,1]))/(max(V[,1])-min(V[,1]))
```

Be careful with brackets and make sure your operation is working correctly. Sometimes it's good to manually check the first few values to be sure. We can use `head(V)` to view the first few entries of the table.

Now let's transform the height variable using standardization. For this, we will use the `sd()` function to calculate the standard deviation.

```
V[,2] <- (V[,2]-mean(V[,2]))/sd(V[,2])
```

> **R Exercise 10** *Use the linear feature scaling technique to get this column and the remaining columns, 3 and 4, to range between 0 and 1.*

2.12.5 Rank-Based Scores

There are three functions in R that can help us with situations where we are interested in the order of arguments in a vector. These are `sort()`, `order()` and `rank()`.

---

[7]All data files can be found at http://www.researchgate.net/publication/306099814_AggWAfit_R_libraryoralternatively, http://aggregationfunctions.wordpress.com/book. These can be saved to your R working directory.

The `sort()` function re-orders a vector into increasing (or non-decreasing) order. So with the vector $\langle 1, 6, 2, 3 \rangle$,

```
sort(c(1,6,2,3))
```

would have an output of 1 2 3 6.

Order, on the other hand, tells us the ordering permutation (the indices from highest to lowest).

```
order(c(1,6,2,3))
```

would have the output 1 3 4 2 because the ordering is $x_1 < x_3 < x_4 < x_2$. If there are ties then the `order()` function will sort them according to the index, i.e. $\langle 3, 2, 3 \rangle$ would be sorted 2 1 3 and not 2 3 1. With both of these functions, we can change to descending order by adding the additional argument `decreasing = TRUE`, i.e.

```
order(c(1,6,2,3), decreasing = TRUE)
```

will produce the output 2 4 3 1.

Rank tells us the relative ranking of the variables. So

```
rank(c(1,6,2,3))
```

will have an output of 1 4 2 3 because the 6 is ranked fourth, the 2 is ranked second etc. The decreasing option is not available for `rank()`, however by using a negative in front of the input vector the opposite ranking will be obtained, i.e.

```
rank(-c(1,6,2,3))
```

would be 4 1 3 2.

The most useful function for us in order to convert our *Sprint* times to rank-scores would hence be the `rank()` function. The following ranks the times and then scales them to the unit interval (the first line retrieves the original times).

```
V[,1] <- original[,1]
V[,1] <- (rank(-V[,1])-1)/(length(V[,1])-1)
```

In this case we used `rank(-V[,1])` because the lowest value would usually be given the rank 1 but we want it to have the highest score. The subtractions of 1 will mean that the worst time will be given a score of zero.

2.12.6 Using `if()` for Cases

Using the `if()` expression requires a careful consideration of the sequence and logic of our function. Let's first consider an example of a piecewise-linear function for values between 0 and 1 that has its join at $(0.5, 0.7)$, i.e. if the input is 0.5, then the output is 0.7. If the input is below 0.5, then it gets increased at the same ratio, and if it is above 0.5, then the ratio of increase drops off so that it still has the output of 1 if the input is 1.

Using Eq. (2.6), we express the function as an equation in the following way

$$f(t) = \begin{cases} 0.7\frac{t}{0.5}, & 0 \leq t < 0.5 \\ 0.7 + 0.3\frac{t-0.5}{0.5}, & 0.5 \leq t \leq 1. \end{cases}$$

As a function in R, we need to create a clause that transforms it using the first equation if the value is less then 0.5, and using the second equation if it is above 0.5. There are a few different ways to do this. The easiest is to use `if(...) {...} else {...}`. Inside the `if()` brackets, we have something like `t < 0.5` or `t >= 0.5` (the latter means greater than or equal to 0.5). It is a logical condition that must evaluate to a single TRUE or FALSE (not NA, and not multiple values). In the first case brackets {...}, we tell the function what to do if the `if()` statement is true, and the second case brackets tells the function what to do otherwise. The following programs our piecewise function above.

```
pw.function <- function(t) {
  if(t < 0.5) {0.7*t/0.5} else {0.7+0.3*(t-0.5)/0.5}
}
```

> **R Exercise 11** *Enter in the function and calculate the outputs for a few values, e.g.* `pw.function(0.1)`, `pw.function(0.6)`, *to see that it makes sense and is working correctly.*

We can repeat this process in nested form to define more cases. In the following, we interpolate the points (0.5, 0.7) and (0.8, 0.9). In this case our intervals in terms of the input cases are [0, 0.5], [0.5, 0.8] and [0.8, 1] respectively.

```
pw.function.2 <- function(t) {
  if(t < 0.5) {0.7*t/0.5}
  else {if(t <0.8) {0.7+0.2*(t-0.5)/0.3}
  else {0.9+0.1*(t-0.8)/0.2} }
}
```

Another way to have a look at our variables and make sure our functions are working correctly is to plot them.

2.12.7 Plotting in Two Variables

If we just want to view a variable to get an idea of the distribution, we can input the vector containing that variable's data. So to plot our original Sprint data.

```
plot(original[,1])
```

This just plots the value in sequence, so along the x-axis is the index of the datum and the y-axis contains its value. It can be easier to see the distribution by first sorting the data. So we can enter

```
plot(sort(original[,1]))
```

Of course, usually histograms are used to plot distributions. So we can also use the following to get a snapshot of our data.

```
hist(original[,1])
```

If we want to plot a single variate function, plotting is reasonably straightforward, however we usually need to create our 'x' values and 'y' values beforehand. To create equispaced x values, the easiest way is to use the seq() command.

So if we want to create a vector of 100 points over the unit interval (starting with 0.01 and going up by 0.01 each entry), we can write seq(0.01,1,length.out=100). If we want to start at 0 (which would give us 101 points), then we can write seq(0,1,length.out=101).
To plot the function $f(t) = t^2$ for these values, we can either use

```
x <- seq(0,1,length.out=101)
plot(x^2)
```

or if we want the x labels to correspond with the data in that variable, we would include these as the first argument and the transformed values as the second argument.

```
x <- seq(0,1,length.out=101)
plot(x,x^2)
```

However, if we want to plot a function like our piecewise function, we need to create the y values separately first. This is because in our cases we used an if-statement, if(t < 0.5), so when we try to input pw.function(x) it will ask whether the vector is less than 0.5, which is not a valid operation. To create the vector of y values, we will also now need to use a repeating operation for().
The for() statement can perform an operation for every entry of a set. This is very useful when we can index our numbers.

The following sequence of operations first creates a vector of y-values (pre-filled with zeros). It uses length(x) so that it will be the same length as our x vector, but we could also just write 100. It then says for all the numbers in 1 to 100 (using the 1:100 vector), each entry in the y vector will be replaced by the piecewise function of the *corresponding* entry in the x vector. The # can be used in R for notes. Anything on the line that comes after the # is "commented out".

```
y <- array(0,length(x))       # 1. create a vector of zeros
for(i in 1:length(x)) {       # 2. perform this operation
  y[i] <- pw.function(xs[i])  #      changing 'i' for the numbers
}                             #      between 1 to 100.
```

We now should be able to plot this piecewise function.

```
plot(x,y)
```

2.12.8 Defining Power Means

We now have all the tools necessary to define our power means. This time we will have two inputs to the function, 'x', which will be a vector of inputs and 'p', which will be the power used.

```
PM <- function(x,p) {      # 1. pre-defining the function inputs
  (mean(x^p))^(1/p)        # 2. our calculation which will also
}                          #    be the output
```

However this function will not work if $p = 0$. So we need to create a special case. For this, we will use the if() command.

```
PM <- function(x,p) {      # 1. pre-defining the function inputs
  if(p == 0) {             # 2. condition for 'if' statement
    prod(x)^(1/length(x))  # 3. what to do when (p==0) is TRUE
  }
  else {
    (mean(x^p))^(1/p)      # 4. what to do when (p==0) is FALSE
  }
}
```

Note here that for '=' conditions, we use a double equals sign '=='. So the possible conditions we can use inside the if() brackets are '==', '<', '>', '<=' and '>='.

R Exercise 12 *Define the power mean as a function in R and check the following.*

Input	Expected output
PM(c(3,2,7),2)	*4.546061*
PM(c(1,0,7),0)	*0*
PM(c(0.28,0.4,0.47),-557)	*0.2805528*
PM(c(0.28,0.4,0.47),-558)	*Inf*

Note that in the last case, 'Inf' should not actually be the result. It's just that at this point, the system cannot tell the difference between one of the transformed inputs and infinity, i.e. if you compare 0.28^(-557) and 0.28^(-558), the latter will be 'Inf' and then all of the operations become absorbed by this value (see [7]). It can be fixed by including a statement in the function definition that if $p < -500$ then the output should be the minimum.

2.13   Practice Questions Using R

1. Suppose you have $\mathbf{x} = \langle 0.3, 0.8, 0.1, 0 \rangle$, Calculate the power mean for the following cases

   (i) $p = 4$
   (ii) $p = 2.5$
   (iii) $p = 0$
   (iv) $p = -3.1$

   and comment on (i.e. compare) the results.
2. Create a 2-variate function for the power mean when $p = 3$ using
   ```
   PM3 <- function(x,y) {...}.
   ```
   i.e. so that it takes the inputs x and y, which will be numbers rather than vectors.
3. Load the data file "wait.service.txt" which has the wait and service times from Sect. 2.11 Question 1, and perform the following.

   (i) Use appropriate scaling techniques so that the two variables both have data given over the same range.
   (ii) Calculate the output of the power mean for $p = -1, p = 0, p = 1$ and $p = 2$.
   (iii) What is the value of service and wait time that has the best aggregated value using each of the values of p?
   (iv) Plot the outputs for each of the functions and compare the results.

4. Load the two data files "comp.acc.txt" and "comp.time.txt" and, either merging them into a single table or keeping them as separate vectors, perform the following.

   (i) Use appropriate scaling techniques so that the two variables both have data given over the same range.
   (ii) Calculate the output of the power mean for $p = -1, p = 0, p = 1$ and $p = 2$.
   (iii) What is the value of accuracy and time that has the best aggregated value using each of the values of p?
   (iv) Plot the outputs for each of the functions and compare the results.

References

1. Beliakov, G., James, S.: Citation based journal ranks: the use of fuzzy measures. Fuzzy Sets Syst. **167**(1), 101–119 (2011)
2. Beliakov, G., Pradera, A., Calvo, T.: Aggregation Functions: A Guide for Practitioners. Springer, Heidelberg (2007)
3. Beliakov, G., Bustince, H., Calvo, T.: A Practical Guide to Averaging Functions. Springer, Berlin/New York (2015)

4. De Veaux, R.D., Velleman, P.F., Bock, D.E.: Stats: Data and Models. Pearson, Essex (2016)
5. Gagolewski, M.: Data Fusion. Theory, Methods and Applications. Institute of Computer Science, Polish Academy of Sciences, Warsaw (2015)
6. Grabisch, M., Marichal, J.-L., Mesiar, R., Pap, E.: Aggregation Functions. Cambridge University press, Cambridge (2009)
7. Lesh, R., Cramer, K., Doerr, H.M., Post, T., Zawojewski, J.S.: Model development sequences. In: Lesh, R.A., Doerr, H.M. (eds.) Beyond Constructivism: Models and Modeling Perspectives on Mathematics Problem Solving, Learning, and Teaching, pp. 35–58. Routledge, New York (2003)
8. R Core Team: R: A language and environment for statistical computing. R Foundation for Statistical Computing, Vienna (2014) http://www.R-project.org/
9. Torra, V., Narukawa, Y.: Modeling Decisions. Information Fusion and Aggregation Operators. Springer, Berlin/Heidelberg (2007)

Chapter 3
Weighted Averaging

In the previous chapters we have looked at some properties of different aggregation functions and learnt some methods for dealing with input variables that differ from one another in terms of scale and distribution. We have seen that aggregation functions can sometimes be described as tending toward either the higher or lower inputs.

Aggregation functions can serve multiple purposes. Sometimes we use them to provide an idea of what is 'normal' for a dataset (i.e. a measure of central tendency) and sometimes we use them to give an overall evaluation. We also noted in Chap. 1 that sometimes the aim of calculating an 'average' is to indicate which repeated value would be equivalent to a number of different values, for example, someone who scores 10 points and then 30 points in basketball over two games has the same average as someone who scores 20 and 20. Another reason we may use calculations like the arithmetic mean is to give a best guess estimation from a number of uncertain or noisy values. For example, if we have multiple temperature sensors within a room, all providing a local temperature with some degree of error, taking an average can give an unbiased estimate of the actual temperature.

As well as being able to use averaging aggregation functions for these purposes, two other key uses are for prediction and analysis, however here we require the concept of aggregation weights. In this chapter we will look at how the importance of sources or variables can be introduced into our mean functions so that particular values will have more influence on the aggregated output.

Assumed Background Concepts

- Interpreting the gradient of linear functions
  What is the gradient of the line $y = 3x - 2$ and what does this represent?

- Multiplication with fractional powers
  How can we simplify $a^{\frac{1}{2}} \times a^{\frac{1}{3}}$?

© Springer International Publishing AG 2016
S. James, *An Introduction to Data Analysis using Aggregation Functions in R*,
DOI 10.1007/978-3-319-46762-7_3

Chapter Objectives

- To be able to apply weighted versions of the averaging functions we have studied so far
- To be able to define and interpret weighting vectors of aggregation functions in context

---

3.1   The Problem in Data: Group Decision Making

Suppose we have candidates being evaluated by three 'judges' for an internship position. Each of the judges provides a score, and these are combined into an overall score for each candidate.

Candidate	Judge 1	Judge 2	Judge 3
Yezi	9	6	4
Jimin	7	7	6
Hyolyn	4	8	8

We could use any of the functions we have discussed so far to give this overall rating—some of them tending toward the higher inputs (in which case Yezi might have an advantage since she has a score of 9 from one of the judges) or lower inputs (in which case Jimin will do better because her lowest score is 6)—however suppose that Judge 1 is the manager and wants her opinion to count for at least as much as the other two judges combined. In this case, the symmetry property satisfied by the means from the previous chapters may not be desirable.

The arithmetic mean for three individuals providing scores x_1, x_2, x_3 is

$$\frac{1}{3}(x_1 + x_2 + x_3),$$

or equivalently we can write

$$\frac{1}{3}x_1 + \frac{1}{3}x_2 + \frac{1}{3}x_3.$$

In order to give the manager (who provides the x_1 score for each of the candidates) more importance, we can instead use the coefficients $\frac{1}{2}, \frac{1}{4}, \frac{1}{4}$ or $0.5, 0.25, 0.25$.

Now the calculation will look like this:

$$\frac{1}{2}x_1 + \frac{1}{4}x_2 + \frac{1}{4}x_3.$$

We refer to this as a **weighted** arithmetic mean and we will denote it WAM(**x**). Aside from symmetry, it exhibits most of the same properties as the 'unweighted' version.

(Thinking Out Loud)
Let's check some of the properties.

Translation invariance—adding something like 3 to the values we would have

$$\frac{1}{2}(x_1 + 3) + \frac{1}{4}(x_2 + 3) + \frac{1}{4}(x_3 + 3),$$

however expanding the brackets gives

$$\frac{1}{2}x_1 + \frac{1}{2}(3) + \frac{1}{4}x_2 + \frac{1}{4}(3) + \frac{1}{4}x_3 + \frac{1}{4}(3),$$

and so regrouping we obtain the same calculation with 3 added,

$$\frac{1}{2}x_1 + \frac{1}{4}x_2 + \frac{1}{4}x_3 + \underbrace{\frac{1}{2}(3) + \frac{1}{4}(3) + \frac{1}{4}(3)}_{=3}.$$

This is the translation invariance property, and we can see that as long as the *three coefficients add to 1*, it should be satisfied.

Monotonicity—we can see that increasing any of the outputs should still make the output increase—only at different rates. Any increase to x_1 will make the function increase at a rate of $1/2$ per unit, while increasing x_2 or x_3 will make the function increase at a rate of $1/4$ per unit (this is the same idea as the gradient of an equation $y = mx + c$).

As it turns out, in addition to satisfying translation invariance and monotonicity, weighted arithmetic means defined with respect to weighting vectors that sum to 1 are also homogeneous, averaging and idempotent. Of course, the order we introduce the variables and multiply them by the coefficients will now affect the output so we no longer have symmetry satisfied. For example, using the weights $\langle 0.5, 0.25, 0.25 \rangle$,

$$\text{WAM}(9, 6, 4) = 4.5 + 1.5 + 1 = 7,$$

is different to

$$\text{WAM}(4, 6, 9) = 2 + 1.5 + 2.25 = 5.75,$$

since in the latter case the 4 has the 50 % weighting attached to it. The choice of weighting vector can hence have a significant impact on decision outcomes. We compare the results that would be obtained with the arithmetic mean and weighted arithmetic mean for the three internship candidates.

Candidate	Judge 1	Judge 2	Judge 3	AM	WAM
Yezi	9	6	4	6.33	7
Jimin	7	7	6	6.67	6.75
Hyolyn	4	8	8	6.67	6

We see that Yezi goes from last place to first place, and Hyolyn from equal first to last. On the other hand, if we made Judge 3 the most important judge, e.g. with multipliers respectively given by $\frac{1}{8}, \frac{3}{8}, \frac{1}{2}$, we could obtain a ranking where Hyolyn wins outright.

There are other reasons why we might need weighted aggregation, e.g. if one of our sources is more reliable (sensors, classifiers, predictors), or in university student evaluation (e.g. for deciding awards) we might want to give more weight to higher level or more relevant subjects. If the inputs are commensurable, the output of the weighted arithmetic can still be interpreted in the same units, however there are some restrictions on the weights we can choose in order to maintain its properties of monotonicity, averaging behavior, etc. Our focus in this chapter will firstly be to understand how the idea of weighting is applied to the means that we have considered so far, after which we will look at how to interpret a function's behavior from its weighting vector. In Chap. 5, we will learn how to find these weights based on data, and so our understanding developed here will enable us to better understand and interpret such processes.

3.2 Background Concepts

We will predominantly be building on the definitions of functions we have seen so far, however it may be useful to be aware of some ideas used in statistical regression.

3.2.1 Regression Parameters

You may have come across or heard about regression models in statistics [4], i.e. where we find a model for a situation in terms of a coefficient and an intercept,

$$y = \beta_1 x + \beta_0.$$

The β_1 term is interpreted as the change in y for every unit change in x, i.e. the gradient of the linear regression line. The β_0 parameter represents the intercept or the value of y when $x = 0$. Regression software is usually used to learn these values from a dataset. We can think of the task performed by such software similarly to finding the m and c terms for the general expression of a line $y = mx + c$.

We are now extending this kind of model for multiple variables. A linear multi-regression model in statistics would be of the form,

$$y = \beta_0 + \beta_1 x_1 + \beta_2 x_2 + \ldots + \beta_n x_n.$$

However the coefficients β_i can be negative, high or low. In the introduction we already alluded to the fact that a negative weight would mean we lose the property of monotonicity, and if the weights don't add to 1 then we will lose things like translation invariance. Furthermore, for aggregation functions defined on $[0, 1]$ or $[0, 10]$ etc., we have boundary conditions $A(0, 0, \ldots, 0) = 0$ and so none of the functions we deal with have an 'intercept' or β_0 term.

Side Note 3.1 *In statistical methods, as well as interpreting the magnitude of β_i (the influence of a variable on the model), it is also necessary to make a judgement about its statistical significance, i.e. whether it could have just come about by chance because we don't have very much data. We will touch on such concepts in Chap. 5, however at that time we will also look at more general methods for evaluating our models.*

3.3 Weighting Vectors

In order for aggregation functions to remain averaging and idempotent (and to satisfy the boundary conditions, e.g. when all the inputs are given over $[0, 1]$, we have $A(1, 1, \ldots, 1) = 1$), we require the weight parameters to add to 1.

Definition (informal) 3.1 (Weighting Vector). We will use the term 'weighting vector' to refer to specific vectors whose components are all non-negative and sum to 1. The length of a weighting vector needs to be the same length as the data input vector.

Definition 3.1 (Weighting Vector). An n-dimensional weighting vector $\mathbf{w} = \langle w_1, w_2, \ldots, w_n \rangle$ is a vector such that $w_i \geq 0$ for all i and

$$\sum_{i=1}^{n} w_i = 1.$$

Side Note 3.2 *There are some circumstances where we might have similar objects where the weights do not add to 1 but the term 'weighting vector' is still used. We can always normalize a vector of non-negative weights $\mathbf{w}$ (i.e. ensure that the weights add to 1) by dividing through by the sum of its components.*

In most statistical regression tasks it's okay to have negative parameters—it just means the variable is negatively correlated with the output. We are restricting ourselves to the special class of aggregation functions, which are monotone increasing, and so we cannot have negative weights. Recall, however that with our approach we would apply a negation function or transform the input in some other way if it were negatively correlated with the output.

A weighted version of the arithmetic mean can be given as follows:

Definition (informal) 3.2 (Weighted Arithmetic Mean). While an arithmetic mean sums all the data and then divides by the number of entries, weighted arithmetic means multiply each input by a coefficient w_i and add the results together. As long as the coefficients are positive and add to 1 (i.e. as long as they are entries of a weighting vector), the output will still be bound between the smallest and highest input.

Definition 3.2 (Weighted Arithmetic Mean). For a given input $\mathbf{x} = \langle x_1, x_2, \ldots, x_n \rangle$, and a weighting vector $\mathbf{w}$, the weighted arithmetic mean is given by

$$\text{WAM}_\mathbf{w}(\mathbf{x}) = \sum_{i=1}^{n} w_i x_i = w_1 x_1 + w_2 x_2 + \cdots + w_n x_n.$$

As with regression parameters, the value of w_i represents how much the output changes with each increase to x_i.

In the case of our manager in the introduction, where we had

$$\mathbf{w} = \left\langle \frac{1}{2}, \frac{1}{4}, \frac{1}{4} \right\rangle, \text{ or } \mathbf{w} = \langle 0.5, 0.25, 0.25 \rangle,$$

each increase of 1 in the x_1 score given by this judge will increase the candidate's score by 0.5, while increases of the same magnitude from either of the other judges will increase the candidate's score by only 0.25. So scores of $\langle 9, 5, 6 \rangle$ respectively from the judges would be as good as scores of $\langle 8, 6, 7 \rangle$, where the drop in the first score needs to be compensated by a larger increase across the other scores.

Example 3.1. Evaluate the weighted arithmetic mean when $\mathbf{w} = \langle 0.2, 0.5, 0.3 \rangle$ and the input $\mathbf{x}$ is $\langle 0.6, 0.7, 0.1 \rangle$.

Solution. Taking the weighting vector into account, our weighted arithmetic mean is explicitly given by

$$0.2x_1 + 0.5x_2 + 0.3x_3.$$

Then since we have $\mathbf{x} = \langle 0.6, 0.7, 0.1 \rangle$, $x_1 = 0.6$, $x_2 = 0.7$ and $x_3 = 0.1$. Our mean is calculated as

$$\text{WAM}_{\mathbf{w}}(0.6, 0.7, 0.1) = 0.2(0.6) + 0.5(0.7) + 0.3(0.1)$$
$$= 0.12 + 0.35 + 0.03 = 0.5.$$

As a side note, if we increase x_2 by 0.1 to 0.8, then the output will go up to 0.55, however if we increase x_1 by 0.1 then the output only goes up to 0.52.

If the weights did not sum to one, e.g. if we had $\mathbf{w} = \langle 0.6, 0.6, 0.6 \rangle$. Then the result of our previous example using $\mathbf{x} = \langle 0.6, 0.7, 0.1 \rangle$ would result in an output of 0.84, which is greater than all the outputs.

3.3.1 Interpreting Weights

The usual interpretation of weights is a set of values that indicate the *importance* of each variable. However weights can be associated with other concepts too.

- In calculating average study scores in university, non-equal weights are sometimes used if different subjects or units contribute differently towards completion of a degree, e.g. year-long subjects compared to semester long, 1 credit point and 2 credit point units. In scholarship or higher degree applications, third year subjects might be weighted more highly.
- Weights could also be used if inputs are reflective of a proportion of the population involved in a decision process. For example, if we were conducting peer evaluation and 5 people recommended a score of 7, 2 people recommended a score of 6 and 3 people recommended a score of 5, then rather than aggregate AM(7, 7, 7, 7, 7, 6, 6, 5, 5, 5) we could calculate a weighted arithmetic mean of $\mathbf{x} = \langle 7, 6, 5 \rangle$ with weights $\mathbf{w} = \langle 0.5, 0.2, 0.3 \rangle$ reflecting the proportion of individuals with each score.

3.3.2 Example: Welfare Functions

In economics research, there is the concept of a 'welfare function'[1].

$$\text{Welfare}_{\mathbf{w}}(\mathbf{x}) = \sum_{i=1}^{n} w_i x_i,$$

where x_i represents the income (or income with respect to the poverty line) of the i-th poorest person. Depending on how many people there are, one scheme for specifying the vector $\mathbf{w}$ is to use the following formula:

$$w_i = \frac{2n + 1 - 2i}{n^2}.$$

For example, if $n = 5$, then $\mathbf{w} = \frac{9}{25}, \frac{7}{25}, \frac{5}{25}, \frac{3}{25}, \frac{1}{25}$. So the 'total welfare' increases more if we increase x_1 (the income of the poorest person) than if we increase x_5.

Notation Note Weight generating function
The equation for finding the w_i based on the value of n will always produce entries that add to 1. In the case of $n = 2$, we would have $w_1 = \frac{2(2)+1-2(1)}{2^2} = \frac{3}{4}$, $w_2 = \frac{2(2)+1-2(2)}{2^2} = \frac{1}{4}$.
In the case of $n = 3$, $w_1 = \frac{2(3)+1-2(1)}{3^2} = \frac{5}{9}$ and so on.

3.4   Weighted Power Means

We can apply weighting vectors in the case of all power means and quasi-arithmetic means too. The weighting convention is analogous to the case of the weighted arithmetic mean.

Definition 3.3 (Weighted Power Mean).   For a given input $\mathbf{x} = \langle x_1, x_2, \ldots, x_n \rangle$ and weighting vector $\mathbf{w}$, the weighted power mean is given by

$$PM_{\mathbf{w},p}(\mathbf{x}) = \left(\sum_{i=1}^{n} w_i x_i^p \right)^{\frac{1}{p}}.$$

For the particular cases of the geometric and harmonic mean we have:

$$PM_{\mathbf{w},0}(\mathbf{x}) = GM_{\mathbf{w}}(\mathbf{x}) = \prod_{i=1}^{n} x_i^{w_i} = x_1^{w_1} \cdot x_2^{w_2} \cdots x_n^{w_n};$$

$$PM_{\mathbf{w},-1}(\mathbf{x}) = HM_{\mathbf{w}}(\mathbf{x}) = \left(\sum_{i=1}^{n} \frac{w_i}{x_i} \right)^{-1} = \frac{1}{\frac{w_1}{x_1} + \frac{w_2}{x_2} + \cdots + \frac{w_n}{x_n}}.$$

Side Note 3.3 *You might notice that the weights seem to be applied differently to the geometric mean. The reason for this becomes clearer if we remember that the geometric mean can also be expressed as a quasi-arithmetic mean via the generator $g(t) = \ln t$. Recall that with logarithms,*

$$b \ln a = \ln(a^b),$$

> *and so when we sum the logarithm of each input and then take the inverse*
> *function at the end, we are left with weights applied as indices rather than*
> *multipliers.*

Notation Note Index notation for functions
We use the notation $PM_{\mathbf{w},p}$ to emphasize the dependence on $\mathbf{w}$ and p. We could also think of these as additional inputs to our function and write $PM(\mathbf{x}, \mathbf{w}, p)$ (and in fact, this is probably the most sensible way to program them in R), however we opt for the sub-script notation to emphasise that once we have $\mathbf{w}$ and p, $PM(\mathbf{x}, \mathbf{w}, p)$ becomes a function that is fixed (so that the values of $\mathbf{w}$ and p are not thought of as being as variable as the values in $\mathbf{x}$).

For means other than the arithmetic mean, we need to be careful in interpreting weights and predicting how the function behaves. We will have two forces at work: firstly, whether the function is influenced more by the higher or lower inputs; and secondly, the weight applied to each value.

Example 3.2. Suppose we are evaluating the attractiveness of a home theatre set up based on the video quality, storage and sound. According to our preference for video quality, we use the weighting vector $\mathbf{w} = \langle 0.5, 0.2, 0.3 \rangle$ to weight the scores in each of the criteria, which are given over the unit interval. We aggregate the scores using the geometric mean. One set up has scores given by $\mathbf{x} = \langle 0.7, 0.6, 0.3 \rangle$ and an overall evaluation of

$$GM_{\mathbf{w}}(\mathbf{x}) = 0.7^{0.5} \times 0.6^{0.2} \times 0.3^{0.3} \approx 0.5264.$$

However we now have the option of spending 100 to increase the quality of one of the criteria—either the video, storage or sound—so that the chosen feature's score goes up by 0.1. Which of the inputs should be increased?

Solution. Since x_1 has the highest weight, we might think it is better to increase the quality of the video.

However, while this would result in an improved score of approximately 0.5627, if we increase the quality of the sound (x_3) by 0.1, we get approximately 0.5738.

In this case, the emphasis the geometric mean places on lower inputs has more influence than the individual weight assigned to w_1, however if the three scores were equally high, it would be better to increase x_1.

The weights still reflect the importance of each variable overall, but we also need to take into account the function's behaviour with respect to high or low inputs.

3.5 Weighted Medians

A median can also be adapted by means of incorporating a weighting vector $\mathbf{w}$ so that the variables are weighted. We will work through an example to give an idea of how it is calculated.

Suppose we have the weighting vector $\mathbf{w} = \langle 0.32, 0.08, 0.20, 0.06, 0.10, 0.24 \rangle$ and the inputs $\mathbf{x} = \langle 0.78, 0.45, 0.03, 0.27, 0.10, 0.45 \rangle$. It will be easier for us if we keep track of these in tabular form.

$\mathbf{w}$	0.32	0.08	0.20	0.06	0.10	0.24
$\mathbf{x}$	0.78	0.46	0.03	0.27	0.10	0.45

Now, as with the standard median, we still re-order our $\mathbf{x}$ vector. Let's do this from lowest to highest, keeping the weights fixed to their respective inputs.

$\mathbf{w}_{(\mathbf{x}\nearrow)}$	0.20	0.10	0.06	0.24	0.08	0.32
$\mathbf{x}\nearrow$	0.03	0.10	0.27	0.45	0.46	0.78

Now we look at the cumulative sum of the weights from left to right and look to the point we have above 50 %. This occurs at the $\mathbf{w}$ entry that is equal to 0.24, since $0.20 + 0.10 + 0.06 = 0.36$ and then $0.36 + 0.24 = 0.6 > 0.5$. So in this case the weighted median output will be 0.45 (the 4th highest input).

On the other hand, if we had the same weighting vector but with the input $\mathbf{x} = \langle 0.62, 0.33, 0.26, 0.11, 0.91, 0.12 \rangle$, then we would have

$\mathbf{w}$	0.32	0.08	0.20	0.06	0.10	0.24
$\mathbf{x}$	0.62	0.33	0.26	0.11	0.91	0.12

and when rearranged in order we get

$\mathbf{w}_{(\mathbf{x}\nearrow)}$	0.06	0.24	0.20	0.08	0.32	0.10
$\mathbf{x}\nearrow$	0.11	0.12	0.26	0.33	0.62	0.91

In this case we have exactly 50 % at the weight equal to 0.20, which is dealt with in different ways. We can either take the half-way point between the two values on the 50 % border

$$\frac{0.26 + 0.33}{2} = 0.295,$$

or we can use the lower value of 0.26 (which is referred to as the 'lower weighted median'), or we can take the higher value of 0.33 (referred to as the 'upper weighted median'). The decision of which value to use as the output will depend on the application and the behavior we want the function to exhibit for 50/50 splits. A more formal definition is as follows.

Definition 3.4 (Weighted (Lower) Median). For an input vector $\mathbf{x} = \langle x_1, x_2, \ldots, x_n \rangle$ and a weighting vector $\mathbf{w}$, the lower weighted median is given by,

$$\text{Med}_\mathbf{w}(\mathbf{x}) = x_{(k)},$$

where the $(\cdot)$ notation indicates that the inputs are first reordered in non-decreasing order, $x_{(1)} \leq x_{(2)} \leq \cdots \leq x_{(n)}$ and k is the index such that

$$\sum_{i=1}^{k-1} w_i < \frac{1}{2} \text{ and } \sum_{i=k+1}^{n} w_i \leq \frac{1}{2}.$$

Example 3.3. Calculate the weighted median when the weighting vectors is given by $\mathbf{w} = \langle 0.1, 0.4, 0.3, 0.2 \rangle$ and the input vector is $\mathbf{x} = \langle 0.3, 0.7, 0.8, 0.2 \rangle$.

Solution. With the input vector, we have $x_4 < x_1 < x_2 < x_3$. The corresponding order for the weights would hence be w_4, w_1, w_2, w_3 or

$$\mathbf{w}_{\mathbf{x}_\nearrow} = \langle 0.2, 0.1, 0.4, 0.3 \rangle.$$

We see that $w_4 + w_1 = 0.3 < 0.5$ and $w_3 = 0.3 \leq 0.5$ and so by definition our output will be the input associated with weight w_2 which is $x_2 = 0.7$.

Notation Note Ordering $\mathbf{w}$ according to $\mathbf{x}_\nearrow$
We have used the notation $\mathbf{w}_{\mathbf{x}_\nearrow}$ to indicate that $\mathbf{w}$ has been reordered according to the ascending order of the associated x values.

3.6 Examples of Other Weighting Conventions

Further examples of how weights can be applied to aggregation can be found in [2, 3, 5, 6, 8]. We will give just a few examples of indices and functions used to summarize data that are similar to weighted power means, however which involve slightly differing weighting conventions.

3.6.1 Simpson's Dominance Index

The concept of 'dominance' (as it is modelled in economics and ecology) is one that can also be considered in the framework of weighted means. We let q_i represent the proportional values based on $\mathbf{x}$, i.e.

$$q_i = \frac{x_i}{\sum_{i=1}^{n} x_i}.$$

For example, if there are four species of birds and their populations in a particular region are given by $\mathbf{x} = \langle 123, 256, 309, 777 \rangle$ then as proportions of the total number of birds, we would have $\mathbf{q} = \langle 0.084, 0.175, 0.211, 0.530 \rangle$. Note that dividing through by the total sum has made this into a weighting vector. Simpson's dominance index, which calculates the degree to which a single species dominates the population, is given by

$$\text{Simp}(\mathbf{q}) = \sum_{i=1}^{n} q_i^2.$$

However we could equivalently write this as

$$\text{WAM}_{\mathbf{w}}(\mathbf{q}) = \sum_{i=1}^{n} w_i q_i,$$

where $w_i = q_i$, i.e. the 'weight' depends on the population.

This function reaches a minimum when all of the inputs are equal, and reaches a maximum of 1 if one of the $q_i = 1$, i.e. there is only one species (since this $q_i = 1$ means that x_i is 100 % of the total).

3.6.2   Entropy

Entropy can be thought of in a similar way. It is expressed:

$$\text{Entropy}(\mathbf{q}) = -\sum_{i=1}^{n} q_i \ln q_i,$$

which is similar to the geometric mean (recall our representation in terms of the quasi-arithmetic mean at the end of the last chapter).

The key difference with these kinds of weighting is that the effective weights will change depending on the inputs, i.e. the smaller the input relative to the remaining inputs, the smaller the weight that will be applied.

> **Side Note 3.4** *This type of weighting convention results in non-monotone behaviour, and so neither Entropy nor Simpson's index can be considered as 'aggregation functions'. For the moment, we will usually restrict ourselves to functions where the weights do not change.*

3.7 The Borda Count

As well as weighting the importance of judges, another use of weighting is in vote-based evaluations. In relation to our first example, this would require each of the selection panel members to rank the candidates rather than give them an overall weighting. Once the ranks are in, each first place vote is worth a certain number of points, each second place is worth a certain number of points (usually less) and so on. The overall winner is then the one who receives the highest number of points.

We can represent this process in our weighted aggregation framework. The entries of our vector **x** would be the total number (or proportion) of i-th placed votes for a particular candidate. The weighted mean then becomes equivalent to tallying up the points. We have

$$\text{Borda}_\omega = \omega_1 x_1 + \omega_2 x_2 + \ldots + \omega_n x_n,$$

where ω_1 is the number of points we give to first place, ω_2 is the number we give to second place and so on. Usually the number of points awarded is obtained from the number of options ranked lower. For example, if an alternative ranked 2nd out of a possible 4, it would receive 2 points since the 3rd and 4th alternatives were ranked below. Applying this kind of weighting to the votes is equivalent to what we obtained in the previous chapter when values were converted into rank-based scores.

Example 3.4. What would be the Borda count for each candidate in the opening example (Sect. 3.1) if the judges' scores were converted to rankings?

Solution. Judge 1's ranking is

$$\text{Yezi} \succ \text{Jimin} \succ \text{Hyolyn},$$

while both Judge 2 and Judge 3 have the ranking

$$\text{Hyolyn} \succ \text{Jimin} \succ \text{Yezi}.$$

Awarding $\omega_1 = 2$ points for first place, $\omega_2 = 1$ point for second place, and $\omega_3 = 0$. The candidate's scores will be:

Yezi	$2(1) + 1(0) + 0 = 2$
Jimin	$2(0) + 3(1) + 0 = 3$
Hyolyn	$2(2) + 1(0) + 0 = 4$

A variant of the Borda count is used in the NBA for judging most valuable player (MVP) of the regular season, defensive player of the year and other awards. In 2015, the top two point scorers were *Kawhi Leonard* and *Draymond Green*. Their votes (from 124 sportswriters and broadcasters are able to vote for three players who they order from first to third) are shown below.

Player	1st place votes	2nd place votes	3rd place votes
Kawhi Leonard	37	41	25
Draymond Green	45	25	17

The weightings were 5 for first place, 3 for second place and 1 for third place. The total number of points obtained by Kawhi Leonard was hence

$$5(37) + 3(41) + 1(25) = 185 + 123 + 25 = 333,$$

while for Draymond Green the point tally gave

$$5(45) + 3(25) + 1(17) = 225 + 75 + 17 = 317.$$

However if an alternative weighting system were used, e.g. with 9 votes for 1st place, 3 votes for second place, and 1 vote for third, we would have Kawhi Leonard with 481 points and Draymond Green with 497 points. It is a common criticism of Borda-type scoring systems that the weighting scheme can have such a big influence on the winner. It can often be arguable whether or not the winning alternative is actually the most preferred.

> **Side Note 3.5** *Note that these Borda count systems do not technically define weighting vectors since they don't add to 1, however they could be normalized to give equivalent results. The outcomes for the NBA awards are equivalent to what would be obtained if we had the weighting vector* $\mathbf{w} = \langle 5/9, 3/9, 1/9 \rangle$.

3.8 Summary of Formulas

Weighting Vector
$\mathbf{w} = \langle w_1, w_2, \ldots, w_n \rangle$ is a vector such that $w_i \geq 0$ for all i and

$$\sum_{i=1}^{n} w_i = 1 \tag{3.1}$$

Weighted Arithmetic Mean

$$\mathrm{WAM}_{\mathbf{w}}(\mathbf{x}) = \sum_{i=1}^{n} w_i x_i = w_1 x_1 + w_2 x_2 + \cdots + w_n x_n \tag{3.2}$$

Weighted Power Mean

$$PM_\mathbf{w}(\mathbf{x}) = \left(\sum_{i=1}^{n} w_i x_i^p \right)^{\frac{1}{p}} \tag{3.3}$$

Weighted Geometric Mean

$$GM_\mathbf{w}(\mathbf{x}) = \prod_{i=1}^{n} x_i^{w_i} \tag{3.4}$$

Weighted (Lower) Median

$$Med_\mathbf{w}(\mathbf{x}) = x_{(k)}, \tag{3.5}$$

where k is the index such that

$$\sum_{i=1}^{k-1} w_i < \frac{1}{2} \text{ and } \sum_{i=k+1}^{n} w_i \leq \frac{1}{2}$$

3.9 Practice Questions

1. There are 6 senior employees at *Lebeau Industries* who will each evaluate candidates for a job position. There are 3 team leaders, 2 managers (who each should have twice as much influence as any of the team leaders) and 1 executive (who should have 1.5 times as much influence as either of the managers).

   (i) What would be an appropriate weighting vector if the scores for each candidate are to be aggregated using a weighted power mean?
   (ii) What would you use as the value for p? Why?

2. Students applying for a foreign scholarship are to be given an overall excellence evaluation based on their performance across 4 subjects: mathematics, computer science, English and Chinese. The scholarship is in China and so Chinese should be weighted highest, however the scores in mathematics and computer science are more important than English scores.

   (i) What would be an appropriate weighting vector if the scores for each candidate are to be aggregated using a geometric mean?
   (ii) What are the advantages of using a geometric mean here rather than an arithmetic mean? What are the disadvantages?

3. Suppose we have three candidates for a job interview and four selection committee members give the following evaluations out of 10.

Candidate	SC 1	SC 2	SC 3	SC 4
A	8	7	9	1
B	3	9	8	5
C	6	4	8	8

(i) What would be the overall score for each candidate using the weighted arithmetic mean if the selection committee are weighted by importance with the vector $\mathbf{w} = \langle 0.3, 0.3, 0.2, 0.2 \rangle$?

(ii) What would be the overall score for each candidate using the geometric mean with the same weighting vector?

(iii) What would be the Borda count score if each first place vote is worth 10 points, each second place vote is worth 5 points, and each third place vote is worth 1 point?

4. Suppose we have three classification algorithms and we test them on 4 datasets. The results (as percentages) and size of each dataset are given as follows:

Classifier	Set 1 (n=89)	Set 2 (n=27)	Set 3 (n=161)	Set 4 (n=68)
Neural network	71	20	87	56
Decision tree	73	29	94	51
k-nearest neighbours	81	64	95	49

(i) Use a weighted arithmetic mean (and justify your choice of weighting vector) to give an overall evaluation of each of the classifiers.

(ii) Using the same weighting vector, use a quadratic mean (a power mean with $p = 2$) to evaluate the methods. Is the relative ranking the same.

(iii) Use a Borda count-based rule in order to rank the different classifiers.

5. The rankings of a set of schools was released with the claim that 15 % of the outcome was based on the school environment, 25 % was based on student performance and 60 % was based on student progress. It was argued that a weighted median should be used to calculate each school's overall score (because medians are more robust to outliers). Do you agree?

6. A weighted power mean is used (with $p = 1/5$) to evaluate different land management plans in terms of biodiversity conservation. The criteria are 1. positive impact to rare species ($w_1 = 0.3$), 2. positive impact to native species ($w_2 = 0.2$), 3. conservation of native vegetation ($w_3 = 0.25$), 4. economic benefit to local (human) population ($w_4 = 0.1$), and 5. cost ($w_5 = 0.15$).

(i) Calculate the score for a management plan with partial evaluations denoted by $\mathbf{x} = \langle 0.6, 0.5, 0.9, 0.7, 0.3 \rangle$.

(ii) If the plan could be slightly improved with respect to one of the criteria, which would you choose?

3.10 R Tutorial

We now need to incorporate weights into our previously obtained functions for means. Our new R [7] functions will have multiple arguments, some of which will be vectors, and some of which will be single values.

3.10.1 Weighted Arithmetic Means

Incorporating weights into the arithmetic mean is quite simple, since you will recall that when we multiply two vectors of the same length, the result is the component-wise products (i.e. $\langle x_1, x_2, x_3 \rangle * \langle y_1, y_2, y_3 \rangle = \langle x_1 y_1, x_2 y_2, x_3 y_3 \rangle$). Once we have this product, taking the sum gives us exactly what we need.

```
WAM <- function(x,w) {
  sum(w*x)
}
```

We can ensure that our weighting vector **w** is normalized (i.e. that the weights add to 1) if we need to by adding an additional line.

```
WAM <- function(x,w) {
  w <- w/sum(w)
  sum(w*x)
}
```

3.10.2 Weighted Power Means

Defining weighted power means is not particularly any more difficult than for the mean. Recall from the previous chapter that we had

```
PM <- function(x,p) {          # 1. pre-define function inputs
  if(p == 0) {                 # 2. condition for 'if' statement
    prod(x)^(1/length(x))      # 3. what to do if (p==0) is TRUE
  }
    else {(mean(x^p))^(1/p)}   # 4. to do when (p==0) is FALSE
}
```

Rather than use the mean, we can use the multiple of our **w** and input vector, which we raise to the power *p*.

```
WPM <- function(x,w,p) {      # 1. pre-defining the inputs
  if(p == 0) {prod(x^w)}      # 2. weighted geom. mean if p=0
  else {sum(w*(x^p))^(1/p)}   # 3. our calculation, which
}                             #    will also be the output
```

3.10.3   Default Values for Functions

We can adapt our power mean function so that we can more easily use it for the geometric mean and arithmetic mean. We do this by including default values, which can be included in the parameter specification. For the default value of *p* we can use 1, while as a default value for a weighting vector, we can use a vector the same length as **x** whose inputs are all $1/n$. So instead of `function(x,w,p)`, we can use

```
function(x,w=array(1/length(x),length(x)),p=1)
```

which means that now we can just input the value for **x**, and are not required to include a weighting vector or value for p in order to obtain an output. If we do include a weighting vector, but not a value for *p*, then we have a weighted arithmetic mean. If we want to leave the weighting vector as is, then we can still change between different kinds of power means, however when we input our function, we need to say that we're using *p*, for example,

```
WPM(c(0.6,0.7,0.8,0.9),p=2)
```

will evaluate the quadratic mean (unweighted) of $\mathbf{x} = \langle 0.6, 0.7, 0.8, 0.9 \rangle$.

R Exercise 13 *Enter in the weighted power mean as a function and verify that you can obtain correct results for different values of p, and* **w**, *and by using the defaults.*

3.10.4   Weighted Median

Programming a weighted median can be a little more complicated than applying weights to other functions, since we need to find the ordered value and check the sums of weights above and below. The following gives one way. The weighting vectors and input vectors are sorted according to **x**. For **x**, this can be achieved using the sort function, however for **w**, we sort it according to the **x** values using `order(x)`. Recall that `order(x)` gives the indices corresponding with the sorting of **x** into increasing order. By listing these inside the square brackets, **w** will be reordered accordingly, e.g. if the ordering of the input vector is 2, 3, 1 (i.e.

the lowest input is the second argument, the next lowest is the third and the highest is the first), then the weights will now be given in the order w_2, w_3, w_1. We will also make use of the cumulative sum function cumsum() and which.max(). The cumulative sum progressively sums the arguments of a vector in order, i.e.

$$\text{cumsum}(\langle 0.1, 0.5, 0.4 \rangle) = \langle 0.1, 0.1 + 0.5, 0.1 + 0.5 + 0.4 \rangle$$

$$= \langle 0.1, 0.6, 1 \rangle.$$

The which.max() function takes a vector as input and returns the index of the largest argument. If there are ties, the index returned is the first one. So which.max($\langle 0.3, 0.1, 0.4, 0.32 \rangle$) = 3 (since the largest argument is the 0.4 which is in the 3rd position) and which.max($\langle 0.5, 0.6, 0.6, 0.6 \rangle$) = 2.

We put these functions together in the following way for the weighted median.

```
Wmed <- function(x,w)  {          # 1. function inputs
  w <- w/sum(w)                   # 2. normalize weights
  w <- w[order(x)]                # 3. sort weights by inputs
  x <- sort(x)                    # 4. sort inputs
  x[which.max(cumsum(w)  >= 0.5)] # 5. return wtd median
}
```

The above calculates the lower weighted median. The last line includes the term cumsum(w) >= 0.5 which returns a vector of TRUE/FALSE values indicating when the cumulative weights are greater than 0.5. Using which.max() here returns the first of these, since all of the TRUE entries are equally 'high', and then this becomes the index for the sorted x vector.

> **R Exercise 14** *Verify the weighted median examples given in the current chapter using the above coded function.*

3.10.5 Borda Counts

Recall that we were able to use the order() and rank() operations to obtain rank-based scores. To define a function that calculates the Borda count, we need to know what form the data will come in. If we have an input vector **x** and each x_i indicates the number or proportion of votes in that position, then we can use a weighted arithmetic mean as we did previously. However if we have input scores that we want to transform into a Borda count scoring system, we will need to do a little more.

If we have rows (representing each candidate or alternative) we can transform the scores they have received from each judge or source (corresponding with the columns) into ranks using our assign function.

We need to replace all columns in the matrix/array with the rank scores. The following will replace the first column with the corresponding rankings.

```
x[,1] <- rank(x[,1])
```

Once this is done, we could then sum the scores in each row (since `rank()` gives the highest value to the highest score), however if we want to use alternative weightings, we need to replace these ranks with those scores.

We can do this using a special indexing command. For a vector, or matrix, `x[x==7]` calls all entries in `x` that are equal to 7. This means we can use the following to convert all rankings of 3 to 10 points.

```
x[x==3] <- 10
```

In order to convert all the rankings, we need to be careful not to assign a score that is equal to one of the unconverted rankings. For example, if we had rankings 1, 2 and 3 and we wanted to assign the respective scores 2, 4, 10, we should do this in the following order.

```
x[x==3] <- 10
x[x==2] <- 4
x[x==1] <- 2
```

Otherwise, after assigning `x[x==1] <- 2`, in the next step both the 2 and original 1 rankings will be changed to 4. An easier way to achieve this, especially when we have multiple replacements and are making replacements for *all* the rankings, is to use `match()`.

```
x <- c(2,4,10)[match(x,c(1,2,3))]
```

In the context of group decision making, the following assumes x is a matrix where each row is a candidate and each column is the set of scores given by one judge. However we need to account for the possibility of tied rankings, which means we will need replacements for 1, 1.5, 2, 2.5, etc., up to the number of ranked alternatives ($2n-1$ in total). The w that needs to be supplied in the following function needs to be the same length. If just the rankings themselves are to be used, we can use w = `seq(1,n,0.5)`.

```
Borda <- function(x,w) {          # 1. input matrix x and w
  total.scores <- array(0,nrow(x))# 2. will hold final scores
  r <- seq(1,nrow(x),0.5)         # 3. the ranks to be matched
  for(j in 1:ncol(x)) {           # 4. convert each column
    x[,j] <- rank(x[,j])          #       to ranks
  }                               #
  for(j in 1:ncol(x)) {           # 5. replace rankings with
    x[,j] <- w[match(x[,j],r)]    #       values in w
  }                               #
  for(i in 1:nrow(x)) {           # 6. add the scores for each
    total.scores[i] <- sum(x[i,]) #       candidate
  }
  total.scores
}
```

3.11   Practice Questions Using R

1. Suppose you have an input vector $\mathbf{x} = \langle 0.88, 0.12, 0.06, 0.46, 0.11 \rangle$ and a weighting vector $\mathbf{w} = \langle 0.26, 0.21, 0.14, 0.39, 0 \rangle$, calculate the weighted power mean for the following cases

   (i) $p = 1$
   (ii) $p = 2$
   (iii) $p = 0$
   (iv) $p = -1$

   and comment on (i.e. compare) the results.

2. Suppose you have the input vector $\mathbf{x} = \langle 0.64, 0.50, 0.35, 0.93, 0 \rangle$ and weighting vector $\mathbf{w} = \langle 0.15, 0.34, 0.16, 0.02, 0.33 \rangle$, calculate the weighted power mean for the following cases

   (i) $p = 1.5$
   (ii) $p = 2.5$
   (iii) $p = 0$
   (iv) $p = -1.5$

   and comment on (i.e. compare) the results.

3. For $\mathbf{x} = \langle 0.37, 0.97, 0.01, 0.84, 0.03 \rangle$ and $\mathbf{w} = \langle 0.24, 0.06, 0.28, 0.14, 0.28 \rangle$, calculate the upper and lower weighted medians.

4. Use the female students volleyball dataset and determine the best two players overall using a Borda count type approach where being ranked first in any category is worth 5 points, second in any category is worth 3 points, and third in any category is worth 1 point (zero points for any lower than 3rd).

References

1. Aristondo, O., García-Lapresta, J.L., Lasso de la Vega, C., Marques Pereira, R.A.: Classical inequality indices, welfare and illfare functions, and the dual decomposition. Fuzzy Sets Syst. **228**, 114–136 (2013)
2. Beliakov, G., Pradera, A., Calvo, T.: Aggregation Functions: A Guide for Practitioners. Springer, Heidelberg (2007)
3. Beliakov, G., Bustince, H., Calvo, T.: A Practical Guide to Averaging Functions. Springer, Berlin/New York (2015)
4. De Veaux, R.D., Velleman, P.F., Bock, D.E.: Stats: Data and Models. Pearson, Essex (2016)
5. Gagolewski, M.: Data Fusion. Theory, Methods and Applications. Institute of Computer Science, Polish Academy of Sciences, Warsaw (2015)
6. Grabisch, M., Marichal, J.-L., Mesiar, R., Pap, E.: Aggregation Functions. Cambridge University Press, Cambridge (2009)
7. R Core Team: R: A language and environment for statistical computing. R Foundation for Statistical Computing, Vienna. http://www.R-project.org/ (2014)
8. Torra, V., Narukawa, Y.: Modeling Decisions. Information Fusion and Aggregation Operators. Springer, Berlin/Heidelberg (2007)

Chapter 4
Averaging with Interaction

When the weighted arithmetic mean is used to build a predictive model, one of the key assumptions being made is that the predictor variables are independent (this is also true for classical linear regression models where the weights needn't add to 1). This justifies the additive nature of the model, with an increase to a single variable affecting the output in the same way regardless of the other values. We have seen that power means with high values for p tend to respond more when higher inputs are increased, while the opposite is true for $p < 1$. In many real world scenarios, whether we are modelling natural phenomena or devising a scoring system, inputs can interact with one another or operate as groups. Some inputs will be correlated, which means that increases to one variable tend to co-occur with increases or decreases in another.

In this topic we will introduce an aggregation function referred to as the **discrete Choquet integral**, which is capable of modelling such relationships. We will also introduce the **ordered weighted averaging** (OWA) function, which is a special case but itself a useful aggregation function that can be used for median-like aggregation (i.e. tending toward central evaluations) or for maximum- and minimum-like behaviour, favouring high and low inputs. There are a number of existing aggregation schemes that can be considered in these frameworks, from evaluating diving and gymnastics in the olympics to the use of robust statistics that alleviates distortion by outliers.

Assumed Background Concepts	
• Correlation How do you interpret a correlation coefficient of $r = -0.73$?	• Sets If $A = \{1, 2, 3\}$ and $B = \{1, 2, 5\}$, what is the intersection $A \cap B$ and union $A \cup B$ of these sets?

© Springer International Publishing AG 2016
S. James, *An Introduction to Data Analysis using Aggregation Functions in R*,
DOI 10.1007/978-3-319-46762-7_4

Chapter Objectives

- To be able to apply OWA weights and calculate how similar a function is to the maximum function
- To introduce the notion of a fuzzy measure and learn how to calculate with the discrete Choquet integral

---

4.1 The Problem in Data: Supplementary Analytics

Social media and search data have enabled the hype surrounding films to be quantified in terms of search statistics, 'likes', 'views' and 'tweets'. A number of studies have been conducted on being able to predict opening weekend box office takings from such data, with varying success [12]. At the pace movies are released, especially with some of the recent franchises (e.g. Marvel superhero films), being able to predict earnings ahead of time allows such production companies to plan future projects and co-ordinate marketing campaigns. A sample of 2016 films are shown in the table below along with some of their social media and search statistics, the number of theatres they were screened in and their earnings in the opening weekend [10, 15].

Film	Facebook (thousands)	YouTube (thousands)	Twitter (thousands)	Google (thousands)	Theatres	Box office (millions)
Deadpool	3470	144,000	228	130	3558	132.4
Dawn of Justice	3880	259,000	1.28	82.8	4242	166.0
Zoolander	1.53	35,000	245	68	3394	13.8
How to be single	232	8.8	43.3	15.1	3343	17.9
My Big Fat Greek Wedding 2	916	4830	16.1	5.21	3133	17.9

It is reasonable to surmise that there might be an approximate relationship between the variables and the total earnings that (with adequate scaling of the variables) could be modelled with aggregation functions. A higher number of *views*, *likes*, *searches* and *theatres* should usually correspond with more money made at the box office. However there are some things about these analytics that we need to take into account.

(Thinking Out Loud)

Views *and* Likes—or—Views *or* likes

Social media platforms are likely to exhibit supplementary relationships. For example, a high number of Facebook likes may be a good indicator of a strong opening weekend at the box office, and this is also true for a high number of Youtube views, however both a high number of likes *and* views does not necessarily correspond with double the earnings. Sometimes different social media outlets appeal to different demographics, and certain demographics may be active across multiple platforms—liking the Facebook page then proceeding to watch the trailer and tweet about it on Twitter. In previous chapters we've introduced functions that could provide values tending toward the maximum (e.g. a power mean with a high value of p or the maximum function itself), however although the social media analytics may be supplementary, other predictors like the number of theatres showing the film might be independent.

This is not easy to deal with, and in fact, all of the functions we have studied so far potentially "double-count" variables that tend to be correlated, or don't give enough weight to criteria that are exchangeable.

The kind of aggregation we are looking for is one that more or less pools together subsets of variables, e.g. something like

$$\text{WAM}(\ \underbrace{\text{Social media}}_{\max(\text{likes, views, tweets})}\ , \text{Google}, \text{Theaters}).$$

This can be accomplished with the discrete Choquet integral, a weighted aggregation function that can model importance and correlation for subsets of inputs. It also includes the ordered weighted averaging (OWA) function as a special case, which is used widely in computational intelligence research. We will look at some notation before considering each function in turn.

4.2 Background Concepts

We have previously introduced the notation $\mathbf{x}_\nearrow$ and $\mathbf{x}_\searrow$ as indicating permutations of the input vector into non-decreasing and non-increasing order respectively. We also sometimes use $\mathbf{x}_\sigma$ to indicate a general permutation of the inputs.

For the median function, we usually use $\mathbf{x}_\nearrow$ to identify the 'middle' input (ordering from lowest to highest), however we could equivalently use $\mathbf{x}_\searrow$ and the result would be the same. We also saw with the weighted median that when we re-ordered the $\mathbf{x}$ vector, we obtained a corresponding permutation of the $\mathbf{w}$ vector which we denoted $\mathbf{w}_{\mathbf{x}_\nearrow}$. In addition to these functions, there are a number of other aggregation frameworks based on the relative ordering of the inputs.

4.2.1 Trimmed and Winsorized Mean

In the field of robust statistics [8, 9], there are a number of means that represent a compromise between the arithmetic mean and the median. While the median is less susceptible to outliers, it also is unaffected by changes to *any* of the inputs that aren't the 1 or 2 central ones.

For a given multiple of 2 denoted by h, the **trimmed mean** discards the highest $\frac{h}{2}$ inputs and the lowest $\frac{h}{2}$ inputs and then takes the average of those remaining. This is precisely the type of averaging that is used in some olympic events.

Example 4.1. A diver receives the scores $\mathbf{x} = \langle 9, 8.8, 9.6, 4.3, 7.6, 8 \rangle$ for one of her dives. What will be her final score?

Solution. The procedure in olympic diving competitions is to remove the highest and lowest evaluations, which is the same as a trimmed mean where $h = 2$. With these scores, we have $\mathbf{x}_\nearrow = \langle 4.3, 7.6, 8.8, 9, 9.6 \rangle$. Discarding the 4.3 (lowest) and 9.6 (highest), her final score will be given by

$$AM(9, 8.8, 7.6, 8) = 8.35.$$

One of the reasons this framework is employed is so that a biased judge (e.g. from the same country as the diver) will not be able to unduly influence the final score. After the reordering of the inputs, such a mean is equivalent to using a weighting vector $\mathbf{w} = \langle 0, \frac{1}{4}, \frac{1}{4}, \frac{1}{4}, \frac{1}{4}, 0 \rangle$.

A similar goal is achieved by the **Winsorized mean**. In this case, the highest and lowest $\frac{h}{2}$ of the inputs are replaced with the next highest/lowest of the inputs.

Example 4.2. Calculate the Winsorized mean for $h = 4$ and $\mathbf{x} = \langle 0.3, 0.8, 0.43, 0.2, 0.33, 0.490.7, 0.4 \rangle$.

Solution. We first reorder the inputs so that we can easily identify the highest and lowest,

$$\mathbf{x}_\nearrow = \langle 0.2, 0.3, 0.33, 0.4, 0.43, 0.49, 0.7, 0.8 \rangle.$$

We now replace the lowest two inputs 0.2 and 0.3 with the next lowest, i.e. $x_{(3)} = 0.33$, and the highest two inputs with 0.49. We then calculate the arithmetic mean.

$$AM(0.33, 0.33, 0.33, 0.4, 0.43, 0.49, 0.49, 0.49) = 0.41125.$$

In this case, after the reordering step, we could consider the aggregation as associated with a weighting vector

$$\mathbf{w} = \left\langle 0, 0, \frac{3}{n}, \frac{1}{n}, \frac{1}{n}, \frac{3}{n}, 0, 0 \right\rangle.$$

We will see that such functions, along with any weighting convention associated with an ordered permutation of the inputs, can be considered as special cases of the ordered weighted averaging (OWA) operator [16].

4.3 Ordered Weighted Averaging

Rather than weights being associated with the source for each input, the OWA allocates a weight depending on the relative ordering of the inputs.

4.3.1 Definition

The OWA is defined as follows.

Definition (informal) 4.1 (Ordered Weighted Averaging (OWA)). To calculate the output for the OWA, we first re-order the inputs *lowest* to *highest*. We then multiply each input by its corresponding weight, with w_1 multiplying by the lowest input, w_2 by the second lowest and so on. These products are then added together.

Definition 4.1 (Ordered Weighted Averaging (OWA)). For a given input $\mathbf{x} = \langle x_1, x_2, \ldots, x_n \rangle$, and a weighting vector $\mathbf{w}$, the OWA is given by

$$OWA_{\mathbf{w}}(\mathbf{x}) = \sum_{i=1}^{n} w_i x_{(i)} = w_1 x_{(1)} + w_2 x_{(2)} + \cdots + w_n x_{(n)}$$

where $(\cdot)$ indicates that the arguments of $\mathbf{x}$ have been rearranged into non-decreasing order, i.e. $x_{(1)} \leq x_{(2)} \leq \cdots \leq x_{(n)}$.

Notation Note Non-decreasing permutations
The notation $x_{(i)}$ is not used universally to signify a non-decreasing permutation of the input vector, and actually it is usually the convention to define the OWA using a non-increasing permutation with w_1 allocated to the highest input. We opt for non-decreasing (ascending) order to avoid confusion and remain consistent with the other functions we cover that involve a permutation of the input vector.

Example 4.3. Calculate the output for the OWA when $\mathbf{x} = \langle 0.3, 0.6, 0.8, 0.2 \rangle$ and $\mathbf{w} = \langle 0.2, 0.4, 0.3, 0.1 \rangle$.

Solution. In this case we have

$$\mathbf{x}_{\nearrow} = \langle 0.2, 0.3, 0.6, 0.8 \rangle,$$

and so to calculate the output with respect to $\mathbf{w}$, we have

$$OWA_{\mathbf{w}}(\mathbf{x}) = 0.2(0.2) + 0.4(0.3) + 0.3(0.6) + 0.1(0.8) = 0.42.$$

4.3.2 Properties

Even though it uses a weighting vector, the OWA operator satisfies the *symmetry* property. This is because it doesn't matter what the initial ordering of our vector is, we always rearrange the values from lowest to highest before we apply the weights. It is also *translation invariant* and *homogenous*, so changes to the scale will not affect relative results for the outputs. Then as with all averaging aggregation functions, it is *monotone* and *idempotent*.

If all the weights are greater than zero, the monotonicity will be *strict* and there will be no absorbent elements, however if some weights are zero then clearly some increases to the inputs may leave the output unchanged. It is only when the OWA corresponds with the minimum and maximum function that it will have absorbent elements (i.e. if it is the maximum, then inputs over $[0, 1]$ will have an absorbent element of 1, while for the minimum over this interval the absorbent element will be 0).

Unlike the weighted arithmetic mean, however, the OWA is not self-dual in general. For a given OWA operator defined with respect to a weighting vector $\mathbf{w}$, the dual will be equal to an OWA with the weights in reverse order. For example, if $\mathbf{w} = \langle 0.3, 0.3, 0.1, 0.1, 0.2 \rangle$, then the weighting vector of the dual function will be $\mathbf{w}^d = \langle 0.2, 0.1, 0.1, 0.3, 0.3 \rangle$.

4.3.3 Special Cases

All the symmetric functions we have introduced so far that require a re-ordering of the inputs from highest to lowest (or lowest to highest) can be expressed as special cases of the OWA, i.e. the median, the maximum and the minimum. The corresponding weighting vectors are given as follows.

$$
\begin{aligned}
&\text{Minimum} &&\mathbf{w} = \langle 1, 0, 0, \ldots, 0 \rangle \\[1ex]
&\text{Maximum} &&\mathbf{w} = \langle 0, 0, \ldots, 0, 1 \rangle \\[1ex]
&\text{Median} &&\mathbf{w} = \langle 0, \ldots, 0, 1, 0, \ldots, 0 \rangle, \quad n = 2k+1 \\
& && \qquad\qquad \underbrace{}_{k \text{ zeros}} \quad \underbrace{}_{k \text{ zeros}} \\[1ex]
& && \mathbf{w} = \langle 0, \ldots, 0, 0.5, 0.5, 0, \ldots, 0 \rangle, \quad n = 2k \\
& && \qquad\; \underbrace{}_{k-1 \text{ zeros}} \qquad\qquad \underbrace{}_{k-1 \text{ zeros}} \\[1ex]
&\text{Trimmed mean} &&\mathbf{w} = \langle 0, \ldots, 0, \tfrac{1}{n-h}, \tfrac{1}{n-h}, \ldots, \tfrac{1}{n-h}, 0, \ldots, 0 \rangle \\
& && \qquad\quad \underbrace{}_{\frac{h}{2} \text{ zeros}} \qquad\qquad\qquad\qquad\quad \underbrace{}_{\frac{h}{2} \text{ zeros}} \\[1ex]
&\text{Winsorized mean} &&\mathbf{w} = \langle 0, \ldots, 0, \tfrac{2+h}{2n}, \tfrac{1}{n}, \ldots, \tfrac{1}{n}, \tfrac{2+h}{2n}, 0, \ldots, 0 \rangle \\
& && \qquad\quad\; \underbrace{}_{\frac{h}{2} \text{ zeros}} \qquad\qquad\qquad\qquad\quad\; \underbrace{}_{\frac{h}{2} \text{ zeros}}
\end{aligned}
$$

In fact, the arithmetic mean (unweighted) is also a special case when $w_i = \frac{1}{n}$ for all i.

4.3.4   Orness

The OWA is useful because it allows us to define functions that graduate between the minimum and maximum function, favoring higher or lower inputs depending on our preference. For example, the weighting vector

$$\mathbf{w} = \langle 0, 0, 0.1, 0.2, 0.3, 0.4 \rangle$$

takes the highest 4 inputs into account and gives the highest one the most influence. We might have a function like this with student tests if we want to take their best test scores into account but also want to reward more than one instance of high performance.

The concept of 'orness' has been introduced [4] to characterize averaging functions (not just the OWA) in terms of how similar to the maximum function they are. The maximum has an orness degree of 1, while the minimum function has an orness degree of 0. The arithmetic mean, which treats high and low inputs equally has an orness of 0.5, while a function like the geometric mean,[1] which is affected more by lower inputs, has an orness level of between 0 and 0.5. We use the term 'orness' because in some extended logical frameworks, the 'OR' operation is modelled by the maximum, while the 'AND' operation is modelled by the minimum.

Side Note 4.1 *For example, with binary inputs,*

$$OR(1, 1) = 1, OR(1, 0) = 1 \ \ and \ \ OR(0, 0) = 0,$$

while

$$AND(1, 1) = 1, AND(1, 0) = 0, AND(0, 0) = 0.$$

Sometimes rather than 'orness', the concept is referred to as the 'degree of disjunction' or 'disjunctive degree'. This again makes reference to OR, which represents a logical disjunction and AND represents a logical conjunction.

Calculating the degree of orness is usually very complicated, involving multivariate integration, however in the case of the OWA operator it can be calculated directly from the weighting vector.

Definition (informal) 4.2 (Orness of OWA). The orness of the OWA function represents how similar it is to the maximum function. Each weight is multiplied by the fraction $\frac{i-1}{n-1}$, e.g. if $n = 4$, then w_1 would be multiplied

---

[1] The orness of the geometric mean changes depending on n. When $n = 2$ the orness is $1/3$. For the harmonic mean, the orness also varies with n, and is approximately 0.2274 when $n = 2$.

by 0, w_2 would be multiplied by $\frac{1}{3}$, w_3 would be multiplied by $\frac{2}{3}$ and w_4 would be multiplied by 1. These products would then be added together to give the orness.

Definition 4.2 (Orness of OWA). For a given weighting vector $\mathbf{w} = \langle w_1, w_2, \ldots, w_n \rangle$, the degree of orness of the associated OWA function is given by

$$\text{orness}(\mathbf{w}) = \sum_{i=1}^{n} w_i \frac{i-1}{n-1}.$$

Note that the first multiplier is always 0, and the last is always 1, so if $\mathbf{w} = \langle 1, 0, 0, \ldots, 0 \rangle$ (i.e. the minimum), it is clear we will have $\text{orness}(\mathbf{w}) = 0$, while the orness of $\mathbf{w} = \langle 0, 0, \ldots, 0, 1 \rangle$ will be 1.

The orness of the arithmetic mean (i.e. when $\mathbf{w} = \langle \frac{1}{n}, \frac{1}{n}, \ldots, \frac{1}{n} \rangle$) is 0.5, however other weighting vectors like $\mathbf{w} = \langle 0, 0.5, 0.5, 0 \rangle$ or $\mathbf{w} = \langle 0.1, 0.2, 0.4, 0.2, 0.1 \rangle$ also have an orness degree of 0.5.

Plots of 2-variate OWA functions with respect to different weighting vectors are shown in Fig. 4.1. In the case of two variables, the orness of the OWA will be equal to w_2.

Example 4.4. Calculate the orness for an OWA with the weighting vector $\mathbf{w} = \langle 0.1, 0.5, 0.4 \rangle$.

Solution. Since the value of $n = 3$, the multipliers for the weights will be

$$\frac{0}{3-1} = 0, \quad \frac{2-1}{3-1} = \frac{1}{2} = 0.5, \quad \frac{3-1}{3-1} = 1$$

respectively. Hence

$$\text{orness}(\langle 0.1, 0.5, 0.4 \rangle) = 0.1(0) + 0.5(0.5) + 0.4(1) = 0.25 + 0.4 = 0.65.$$

Since this is greater than 0.5, we would say that this particular OWA tends more toward higher inputs.

4.3.5   Defining Weighting Vectors

With power means, the choice of weights is more straightforward, with the importance of a variable or criterion i corresponding with the weight w_i. With OWA weighting vectors, we usually interpret weights in terms of favoring high, low, or central inputs, however some techniques also exist for defining weighting vectors that model natural language.

This has been a particular focus of research around natural language processing and the concept of 'computing with words'. The typical scenario here is that each of the inputs represents the degree of truth regarding some statement, and then the OWA weights represent a quantifier. For example, consider the statement,

Most of the clients were <u>happy with our service.</u>

We may have five clients who completed a survey and they each responded to the statement, "I was happy with my service experience." as indicated below. The options they could choose from were {Extremely happy, very happy, somewhat happy, neither happy nor unhappy, somewhat unhappy, not happy, very unhappy} and each of these was associated with a numerical score out of 1.

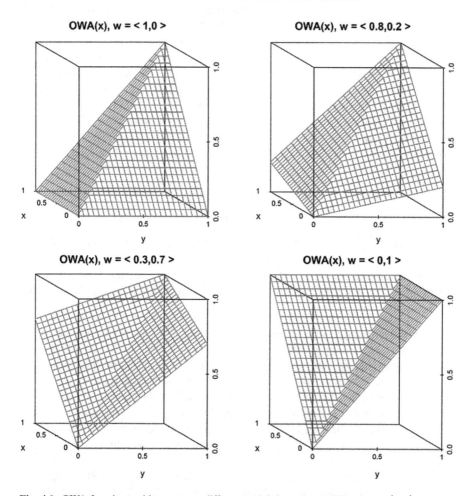

Fig. 4.1 OWA functions with respect to different weighting vectors. When $\mathbf{w} = \langle 1, 0 \rangle$ we have the minimum function and $\mathbf{w} = \langle 0, 1 \rangle$ gives the maximum. The weighting vector $\mathbf{w} = \langle 0.8, 0.2 \rangle$ is closer to the minimum (orness of 0.2) while $\mathbf{w} = \langle 0.3, 0.7 \rangle$ (orness of 0.7) is closer to the maximum

Client	1	2	3	4	5
Response	Very Happy	Very Happy	Somewhat Happy	Not Happy	Extremely Happy
Score	0.8	0.8	0.6	0.3	1

Our aim is to aggregate all of these scores into an overall score telling us how "true" the statement is. We hence need to think of a weighting vector that captures the idea of 'most' when we say "most of the clients".

(Thinking Out Loud)
Let's consider the weighting vectors at either side of the scale. If we have $\mathbf{w} = \langle 0, \ldots, 0, 1 \rangle$, i.e. the maximum, then this just reproduces the best score. It would actually be closer to saying "at least one person was satisfied with our service". On the other hand, the weighting vector $\mathbf{w} = \langle 1, 0, \ldots, 0 \rangle$, which is the minimum, would be equivalent to saying "*everyone* was satisfied" at least to a given degree. We therefore need something in between. A weighting vector like $\mathbf{w} = \langle 0, 0, 1, 0, 0 \rangle$, which is the median, tells us how much at least half of the respondents were satisfied, while $\mathbf{w} = \langle 0, 1, 0, 0, 0 \rangle$, which gives the second lowest score, would mean that at least 80 % were satisfied. This is probably closer to the idea of "most". One option might be to aggregate the highest 4 inputs, with more weight allocated to the lower to middle values, e.g. $\mathbf{w} = \langle 0, 0.4, 0.4, 0.2, 0 \rangle$.

Aggregating with this weighting vector, the truth of the statement would be

$$\text{OWA}(\mathbf{x}) = 0.3(0) + 0.6(0.4) + 0.8(0.4) + 0.8(0.2) + 1(0) = 0.72.$$

On the other hand, the truth of the statement, "everyone was happy with our service" would be the lowest score, which is 0.3.

In this example, we only had 5 responses, however we can also define the weighting vector that means 'most' for 100 or even 1000s of inputs. We do this by means of quantifier functions.

Definition (informal) 4.3 (Quantifier Function). A quantifier function in the context of defining OWA weights is a function of a single variable that is increasing and defined over the interval 0 to 1 with an output that is also from 0 to 1. These can often be constructed as piecewise-linear functions (even discontinuous functions are okay). Once we have the quantifier function, the weights of the OWA are determined by taking the differences in the quantifier output when we take equal steps across the domain. For example, if our quantifier was $Q(t) = t^2$ and we were defining a weighting vector for $n = 4$, then we would evaluate the quantifier at $0.25, 0.5, 0.75$ and 1. Since $f(0.25) = 0.0625$, the value of w_1 would be 0.0625. Then $Q(0.5) = 0.25$, so the difference between this and our first point is $0.25 - 0.0625 = 0.1875$, which becomes our value for w_2. Continuing in the same manner, we have $w_3 = 0.75^2 - 0.5^2 = 0.3125$ and $w_4 = 1 - 0.75^2 = 0.4375$.

Definition 4.3 (Quantifier Function).   A quantifier function Q is an increasing function (strict or not strict) satisfying $Q(0) = 0$ and $Q(1) = 1$. For a given n, the OWA weights associated with the quantifier can be calculated for each i using

$$w_i = Q\left(\frac{i}{n}\right) - Q\left(\frac{i-1}{n}\right).$$

Example 4.5. Find the weighting vector corresponding with the quantifier function $Q(t) = \sqrt{t}$ for $n = 5$.

Solution. Using the definition, we need to evaluate the quantifier at $0.2, 0.4,$ $0.6, 0.8$ and 1 (i.e. $\frac{1}{5}, \frac{2}{5}$, and so on). Therefore we have (to 4 decimal places),

$$w_1 = \sqrt{0.2} - \sqrt{0} = 0.4472, \quad w_2 = \sqrt{0.4} - \sqrt{0.2} = 0.1852,$$

$$w_3 = \sqrt{0.6} - \sqrt{0.4} = 0.1421, \quad w_4 = \sqrt{0.8} - \sqrt{0.6} = 0.1198,$$

$$w_5 = \sqrt{1} - \sqrt{0.8} = 0.1056.$$

Note that these add to 0.9999 so we would need to make an adjustment so that the values add to 1 or use the non-rounded numbers if we were then to evaluate outputs for different input vectors.

We provide some common natural language quantifiers and example quantifier functions that could be used.

$$\text{All} \qquad Q(t) = \begin{cases} 0, t = 0, \\ 1, t > 0. \end{cases}$$

$$\text{At least 1} \quad Q(t) = \begin{cases} 0, t < 1, \\ 1, t = 1. \end{cases}$$

$$\text{Majority} \quad Q(t) = \begin{cases} 0, t < 0.5, \\ 1, t \geq 0.5. \end{cases}$$

$$\text{Some} \qquad Q(t) = \begin{cases} 0, & t < 0.5, \\ \frac{t-0.5}{0.3}, & 0.5 \leq t < 0.8, \\ 1, & t \geq 0.8 \end{cases}$$

$$\text{Most} \qquad Q(t) = \begin{cases} 0, & t < 0.2, \\ \frac{t-0.2}{0.3}, & 0.2 \leq t < 0.5, \\ 1, & t \geq 0.5 \end{cases}$$

4.4 The Choquet Integral

So far we have looked at applying weights to arguments either depending on their source (weighted power means) or depending on their relative size (the OWA). We now turn to a very useful (but a somewhat complex) operator that can take both into account.

The essential idea is that we allocate the weight to different *subsets* of arguments. We see many examples day to day where this makes intuitive sense. In medical diagnosis, symptoms occurring together may present a strong case for a particular condition, whereas by themselves they may mean very little. The output of workers can sometimes by greater when they work together than what they would contribute working alone, while in some cases, their skills may not be complementary and the output becomes sub-additive.

One concept that allows these kinds of interaction to be represented is known as a 'fuzzy measure' or 'capacity'.

4.4.1 Fuzzy Measures

A fuzzy measure [1, 2, 5, 7, 14] allocates a weight to a **set** of inputs, with the property that the full set has a value of 1, the empty set has a value of zero, and if we add an element to the set, we can't decrease the measure, i.e. we have a kind of monotonicity in terms of adding elements to the set.

We will use the notation v to denote a fuzzy measure, and $v(S)$ to denote the value of a subset S according to that measure. We will use $\{1 : n\}$ to refer to the set $\{1, 2, 3, \ldots, n\}$. Note that for a given n, there are 2^n subsets (including the empty set and whole set). For example, in the case of $\{1 : 3\}$ the subsets are $\emptyset, \{1\}, \{2\}, \{3\}, \{1, 2\}, \{1, 3\}, \{2, 3\}$ and $\{1, 2, 3\}$.

Definition (informal) 4.4 (Fuzzy Measure). A fuzzy measure allocates a value between 0 and 1 to each of the subsets of $\{1 : n\}$. To be considered a fuzzy measure, two conditions must be satisfied. Firstly, the value of the whole set ($\{1 : n\}$) is 1 and the value of the empty set ($\emptyset$, a set without any members) is 0. Secondly, it must satisfy monotonicity with respect to adding new elements to a subset. For example, if S is the set consisting of the three members, 1, 2 and 4, and T includes $1, 2, 4$ as well but also includes 6, then $v(T)$ should not be less than $v(S)$.

Definition 4.4 (Fuzzy Measure). For a given finite set $\{1 : n\} = \{1, 2, \ldots, n\}$, a fuzzy measure is a set function v defined for all $S \subseteq \{1 : n\}$ such that $v(\emptyset) = 0$, $v(\{1 : n\}) = 1$ and $S \subseteq T$ implies $v(S) \leq v(T)$.

> **Notation Note** Set notation
> It is common to use the brackets $\{\ldots\}$ when we are referring to discrete sets and need to explicitly list the elements. In our case, we will only be dealing with finite sets (i.e. since we will usually have an input vector of n variables, $\mathbf{x} = \langle x_1, x_2, \ldots, x_n \rangle$). The $S \subseteq T$ can be read "S is a subset of T", and means that all of the members of S are also members of T, however the converse does not necessarily hold. The set $S = \{1, 2, 6\}$ is not a subset of $T = \{1, 2, 4, 5\}$ because 6 is not in T, but it *is* a subset of $T = \{1, 2, 5, 6\}$.

Here is an example of a set of fuzzy measure weights when there are 3 members/elements in the full set.

$$v(\{1, 2, 3\}) = 1$$

$$v(\{1, 2\}) = 0.9 \quad v(\{1, 3\}) = 0.8 \quad v(\{2, 3\}) = 0.7$$

$$v(\{1\}) = 0.3 \quad v(\{2\}) = 0.3 \quad v(\{3\}) = 0.7$$

$$v(\emptyset) = 0$$

In the case of three variables, the fuzzy measure v is defined for eight different subsets. Two of these are the empty set ($\emptyset$) and the full set, which always have respective values of 0 and 1.

Looking at the other subsets in this example, we see that $v(\{1, 3\}) = 0.8$, even though the first variable has a value of 0.3 and the third variable has a value of 0.7, i.e. the weight of these two inputs together is not worth the same as their individual values added together. To give an idea of what this means in practice, an input of $\mathbf{x} = \langle 1, 0, 1 \rangle$ used with this fuzzy measure would give an output of 0.8, i.e. the value of $v(\{1, 3\})$ since the first and third inputs are 1. The inputs $\mathbf{x} = \langle 1, 0, 0 \rangle$ and $\mathbf{x} = \langle 0, 0, 1 \rangle$ would have outputs of 0.3 and 0.7 respectively (being associated with $v(\{1\})$ and $v(\{3\})$), so there is some kind of redundancy between these two variables.

On the other hand, the first two variables together have a complementary effect. Since by themselves they are only worth 0.3, but together they are worth 0.9. The following is a commonly used example [11].

> *Example 4.6.* Suppose there are three workers and their output depending on the team is determined by the fuzzy measure shown previously. The interpretation of the fuzzy measure values is that if the three work together, they can achieve 100 % of a project per day, if workers 1 and 2 work together, they can finish 90 % of a project per day, worker 1 working by him/herself can finish 30 % of a project per day, and so on. The proportion of a day's work the three are willing to put in is denoted by $\mathbf{x} = \langle 0.3, 0.6, 0.4 \rangle$. If we have a

constraint that all workers have to start at the same time (and can't return once
we send them home), the amount of work achieved is given by,

$$\underbrace{0.3v(\{1,2,3\})}_{\text{the 3 working together}} + \underbrace{(0.4 - 0.3)v(\{2,3\})}_{\text{after worker 1 went home}} + \underbrace{(0.6 - 0.4)v(\{2\})}_{\text{after worker 3 went home}}$$

$$= 0.3(1) + 0.1(0.7) + 0.2(0.3) = 0.43.$$

This calculation with respect to a fuzzy measure is referred to as the discrete
Choquet integral [3]. We will provide the definition and then an example of how it
can be calculated in general.

4.4.2   Definition

The discrete Choquet integral behaves similarly to a weighted arithmetic mean,
however the weights that are applied change depending on the order of the inputs.
We will define the Choquet integral according to a non-decreasing permutation of
the inputs.

Definition 4.5. For a given set of inputs $\mathbf{x} = \langle x_1, x_2, \ldots, x_n \rangle$, and a fuzzy
measure v, the Choquet integral is given by

$$C_v(\mathbf{x}) = \sum_{i=1}^{n} x_{(i)} \Big(v(\{(i) : (n)\}) - v(\{(i+1) : (n)\}) \Big)$$

where the notation $x_{(i)}$ means the inputs are re-arranged in non-decreasing
order and $\{(i) : (n)\}$ refers to the set of (ordered) inputs from the (i)-th smallest
to the largest and $\{(i+1) : (n)\}$ is the set of the $(i+1)$-th smallest to the largest.

Notation Note Alternative representations of C_v
Before we actually calculated the result using an equivalent formula

$$C_v(\mathbf{x}) = \sum_{i=1}^{n} (x_{(i)} - x_{(i-1)}) v(\{(i) : (n)\})$$

with the convention that $x_{(0)} = 0$. However from herein we will show our calculations using
the formula in the definition.

For each term $x_{(i)}$, we have a weight that is calculated using the fuzzy measure. The
measure of the set of all elements greater than $x_{(i)}$ is subtracted from the set of the same
elements, however with (i) included. For example, if we had $\mathbf{x} = \langle 0.3, 0.6, 0.2, 0.8 \rangle$ then
the weight associated with the element $x_1 = 0.3$ would be

$$v(\{1, 2, 4\}) - v(\{2, 4\}).$$

In fact we technically don't need to re-order our input vector, we just need to know the
ordering so that we know which subsets to use to calculate the weight.

4.4.3 Special Cases

The discrete Choquet integral includes both the WAM and the OWA as special cases (and therefore all the special cases of the OWA too).

> (WAM) When the fuzzy measure is such that
>
> $$v(S) = \sum_{i \in S} v(\{i\}),$$
>
> i.e. the value of any subset is just the combined total of all the values of its singletons, the resulting Choquet integral will be equivalent to a weighted arithmetic mean where $w_i = v(\{i\})$ for each i. A fuzzy measure with this property is also referred to as a probability measure.
>
> (OWA) The Choquet integral will be equivalent to an OWA when the fuzzy measure is such that all subsets of the same size have the same value, i.e. $v(\{1\}) = v(\{2\}) = \cdots = v(\{n\})$ and $v(\{1,2\}) = v(\{1,3\})$, and so on. When this is the case, the weights of the corresponding OWA can be calculated using $w_i = v(S_{|S|=n-i+1}) - v(S_{|S|=n-i})$, where $S_{|S|=a}$ means that the cardinality is a. For example, if $n = 6$ then the weight for w_2 will be the fuzzy measure value of the subsets with cardinality 4 $v(S_{|S|=4})$ subtracted from the value of subsets with cardinality 5 $v(S_{|S|=5})$.

Plots of the Choquet integral with respect to some example fuzzy measures are shown in Fig. 4.2.

4.4.4 Calculation

The most straightforward way of calculating the output is to calculate the weights from the fuzzy measure based on the input vector and then apply these in the same way we would a weighted arithmetic mean.

Let's work through a relatively simple example with $\mathbf{x} = \langle 0.3, 0.7, 0.2 \rangle$ and the values of the fuzzy measure given as follows.

$$v(\{1,2,3\}) = 1$$

$$v(\{1,2\}) = 0.4 \quad v(\{1,3\}) = 0.9 \quad v(\{2,3\}) = 0.5$$

$$v(\{1\}) = 0.3 \quad v(\{2\}) = 0.4 \quad v(\{3\}) = 0.1$$

$$v(\emptyset) = 0$$

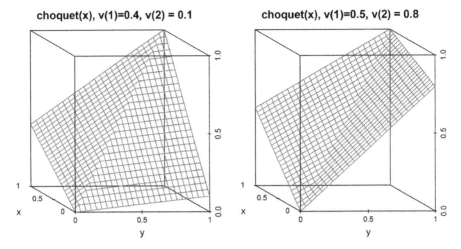

Fig. 4.2 Plots of the Choquet integral for two inputs $C_v(x, y)$. The fuzzy measure on the left has $v(\{1\}) = 0.4, v(\{2\}) = 0.1$ (and as is always the case for two inputs, $v(\emptyset) = 0, v(\{1, 2\}) = 1$). This means both of the inputs need to be high for a high output (inputs complement one another) and the function behaves more like the minimum. On the *right* we have a fuzzy measure where the inputs are somewhat redundant, $v(\{1\}) = 0.5, v(\{2\}) = 0.8$. Only one of the inputs needs to be high for a relatively high output and the function is closer to the maximum

The ordering of **x** is $x_3 \prec x_1 \prec x_2$ (we use $\prec$, which just means "is ordered before", instead of $<$ in case there are ties). From this, let's compile a table of the subsets associated with each value that are used to determine the weights.

Input	$v(\{(i) : (n)\})$ fuzzy measure of i and all j such that $x_j \succ x_i$	$v(\{(i+1) : (n)\})$ fuzzy measure of all j such that $x_j \succ x_i$	Calculated weight
x_1	$v(\{1, 2\}) = 0.4$	$v(\{2\}) = 0.4$	$v(\{1, 2\}) - v(\{2\})$ $0.4 - 0.4 = 0$
x_2	$v(\{2\}) = 0.4$	$v(\emptyset) = 0$	$v(\{2\}) - v(\emptyset)$ $0.4 - 0 = 0.4$
x_3	$v(\{1, 2, 3\}) = 1$	$v(\{1, 2\}) = 0.4$	$v(\{1, 2, 3\}) - v(\{1, 2\})$ $1 - 0.4 = 0.6$

So the final calculation will be

$$0.3(0) + 0.7(0.4) + 0.2(0.6) = 0 + 0.28 + 0.12 = 0.4$$

Side Note 4.2 *If we do have ties, e.g., if we had* **x** $= \langle 0.6, 0.3, 0.3 \rangle$, *then the ordering* $x_2 \prec x_3 \prec x_1$ *would mean that for* x_2 *we use* $v(\{1, 2, 3\}) - v(\{1, 3\})$ *and for* x_3 *we use* $v(\{1, 3\}) - v(\{1\})$. *Alternatively, if we use the ordering* $x_3 \prec x_2 \prec x_1$, *we use* $v(\{1, 2, 3\}) - v(\{1, 2\})$ *for* x_3 *and* $v(\{1, 2\}) - v(\{1\})$ *for* x_2.

We can use the same method when we have more variables. For example, if we have $\mathbf{x} = \langle 0.3, 0.6, 0.7, 0.4, 0.1 \rangle$, we would calculate the weights according to the following table.

Input	$v(\{(i) : (n)\})$	$v(\{(i+1) : (n)\})$	Calculated weight
x_1	$v(\{1, 2, 3, 4\})$	$v(\{2, 3, 4\})$	$v(\{1, 2, 3, 4\}) - v(\{2, 3, 4\})$
x_2	$v(\{2, 3\})$	$v(\{3\})$	$v(\{2, 3\}) - v(\{3\})$
x_3	$v(\{3\})$	$v(\emptyset)$	$v(\{3\}) - v(\emptyset)\ (= v(\{3\}))$
x_4	$v(\{2, 3, 4\})$	$v(\{2, 3\})$	$v(\{2, 3, 4\}) - v(\{2, 3\})$
x_5	$v(\{1, 2, 3, 4, 5\})$	$v(\{1, 2, 3, 4\})$	$v(\{1, 2, 3, 4, 5\}) - v(\{1, 2, 3, 4\})$

The calculations in the last column will always result in a weighting vector with weights that sum to 1.

4.4.5 Examples

Another often given example in the literature is the problem of evaluating students for a scholarship based on their scores in Mathematics, Physics and English [6].

Suppose the students' scores are given as follows.

Student	Maths	Physics	English
Sachin	18	16	10
Courtney	10	12	18
Wasim	14	15	15

One thing that we want to account for is that students who are good at maths are often also good at physics, so in combining their scores, we don't necessarily want to 'double count' this ability. On the other hand, we want the importance of maths and physics to be higher than English. So we might say that Wasim is the best candidate, since he is strong across the board. However there is no weighted mean consistent with our preferences that would give this result. If mathematics and physics are weighted even only slightly higher than English, e.g. with $\mathbf{w} = \langle 0.34, 0.34, 0.32 \rangle$ then Sachin gets a score of 14.76 and Wasim will receive 14.66.

Fuzzy measures make such preferences possible. We can allocate the values as follows.

$$v(\text{maths}) = v(\text{physics}) = 0.45$$

$$v(\text{English}) = 0.3$$

$$v(\text{maths, physics}) = 0.5$$

$$v(\text{physics, English}) = v(\text{maths, English}) = 0.9$$

Then obviously we will have 0 and 1 for the empty set and full set respectively. Now the calculations will be:

For Sachin, we have

$$10(1 - 0.5) + 16(0.5 - 0.45) + 18(0.45) = 13.9.$$

For Courtney, we have

$$10(1 - 0.9) + 12(0.9 - 0.3) + 18(0.3) = 13.6.$$

Finally for Wasim, we have

$$14(1 - 0.9) + 15(0.9 - 0.3) + 15(0.3) = 14.9.$$

So we were able to assign higher value to maths and physics, but still reward all-round ability.

Returning to our opening example of predicting Box Office takings, a discrete Choquet integral could be used to aggregate the predictors such that the collective importance of the online analytics, i.e. YouTube views, tweets, etc., could be worth 80 % of the weight and the number of theatres variable worth 20 %, however we could also allocate high individual weights to the indices. For example, we can assign

$$v(\text{Facebook}) = 0.6, \quad v(\text{YouTube}) = 0.7,$$
$$v(\text{Twitter}) = 0.55, \quad v(\text{Google}) = 0.65,$$

for the singleton values, but then use

$$v(\text{Facebook, Youtube, Twitter, Google}) = 0.8$$

for the weight allocated to the subset (and values in between 0.7 and 0.8 for the pairs and triples).

The prediction would then essentially combine the highest scoring analytics predictor along with the score according to the number of theatres screening the film. This helps alleviate the problem of double counting.

Example 4.7. Calculate the discrete Choquet integral of the input vector $\mathbf{x} = \langle 0.8, 0.3, 0.4 \rangle$, where the fuzzy measure values are given by

$$v(\{1, 2, 3\}) = 1, \quad v(\{1, 2\}) = v(\{1, 3\}) = 0.6, \quad v(\{2, 3\}) = 0.9,$$
$$v(\{1\}) = 0.4, \quad v(\{2\}) = v(\{3\}) = 0.1.$$

Solution. Rearranging the vector we have $\mathbf{x}_\nearrow = \langle 0.3, 0.4, 0.8 \rangle$. We can calculate the output directly as

$$C_v(\mathbf{x}) = 0.3(v(\{1, 2, 3\}) - v(\{1, 3\})) + 0.4(v(\{1, 3\}) - v(\{1\})) + 0.8v(\{1\})$$

$$= 0.3(1 - 0.6) + 0.4(0.6 - 0.4) + 0.8(0.4)$$

$$= 0.3(0.4) + 0.4(0.2) + 0.8(0.4) = 0.12 + 0.08 + 0.32 = 0.52$$

Example 4.8. Show that the 'counting measure', i.e. the set function $v(S) = \frac{|S|}{n}$, defines a fuzzy measure for any given n. (Note that $|S|$ means the cardinality (number of elements) in S)

Solution. First we can check the boundary conditions. When S is the empty set, we will have $v(S) = \frac{0}{n}$. On the other hand, when S is the full set, we have $S = \{1 : n\}$ and therefore $|S| = n$ so $v(S) = \frac{n}{n} = 1$. We then need to make sure that adding elements to a subset does not decrease $v(S)$. Since it is based on cardinality, the size will always increase by $\frac{1}{n}$ whenever a new element is added, so it is clear that the monotonicity with respect to subsets will hold.

4.4.6 Properties

Similar to the OWA and WAM, the Choquet integral is *translation invariant* and *homogeneous*, however in general it will not be symmetric. It is an averaging aggregation function, and as such is *monotone* and *idempotent*.

The dual of a given discrete Choquet integral is also a Choquet integral, defined with respect to the dual fuzzy measure. The dual fuzzy measure can be calculated where each value is the measure of the complement set subtracted from 1. The 'complement' set is all the members in the whole set that aren't in S. So if $n = 5$ then the dual fuzzy measure value of the set $\{1, 4\}$ would be $1 - v(\{2, 3, 5\})$.

As with the OWA, the orness of the Choquet integral can be calculated directly from the values of the associated fuzzy measure. The calculation can be expressed

$$\text{orness}(v) = \frac{1}{n-1} \sum_{S \subset \{1:n\}} \frac{(n - |S|)! |S|!}{n!} v(S).$$

Notation Note Set cardinality and factorial (!)

Recall that the notation $|S|$ means the *size* of the subset, i.e. how many elements there are in it. So if $S = \{1, 4, 6, 7\}$ then $|S| = 4$. We also use the subset symbol $\subset$ indicating we take all of the proper subsets of the whole set (not including the whole set $S = \{1 : n\}$ itself). We then have the factorial notation, "!", which means we multiply all values up to that number.

We have $3! = 1 \times 2 \times 3 = 6$ and $4! = 1 \times 2 \times 3 \times 4$ while by convention, $0! = 1$. For example, if we have $n = 3$, and the subset $S = \{1, 3\}$, then the associated term in the calculation would be

$$\frac{(3-2)!2!}{3!} v(\{1, 3\}) = \frac{2}{6} v(\{1, 3\}) = \frac{1}{3} v(\{1, 3\}).$$

As well as using the orness to interpret how much a discrete Choquet integral tends toward high or low values, we also have a calculation that indicates the average importance of each variable, which we can interpret similarly to the weights of a weighted arithmetic mean.

4.4.7   The Shapley Value

When n gets large, the number of subsets grows exponentially, doubling with each new variable so that for $n = 3$ we have 8 subsets, for $n = 4$ we have 16 and so on. This means we can have fuzzy measures that are very difficult to interpret in terms of the importance of variables and subsets of variables.

The Shapley value hence has been used to measure the average importance of each variable. It is denoted by the vector

$$\boldsymbol{\phi} = \langle \phi_1, \phi_2, \ldots, \phi_n \rangle,$$

and can be calculated directly from v.

Definition 4.6. For a given fuzzy measure v, the Shapley index for each i in $\{1, 2, \ldots, n\}$ is given by

$$\phi(i) = \sum_{S \subseteq \{1:n\}\setminus\{i\}} \frac{(n - |S| - 1)!|S|!}{n!} \big(v(S \cup \{i\}) - v(S)\big).$$

Notation Note Subset notation
You will note the similarities between this calculation and the one used for orness. The $S \subset \{1 : n\}\setminus\{i\}$ means that for a given i, we take all the subsets that don't include i. The $\cup$ symbol means set union, and here it is used to add i to the set. So as an example, if we were calculating the Shapley index for $i = 2$, then when $n = 4$ the term associated with the subset $S = \{1, 4\}$ would be

$$\frac{(4 - 2 - 1)!2!}{4!} \big(v(\{1, 2, 4\}) - v(\{1, 4\})\big) = \frac{1}{12} \big(v(\{1, 2, 4\}) - v(\{1, 4\})\big)$$

As with a weighting vector, the Shapley value always has Shapley indices that are greater than or equal to zero and add to 1.

There is another calculation that is sometimes used instead of the Shapley index called the Banzhaf index. It has a similar calculation, however with each of the $v(S \cup \{i\}) - v(S)$ equally weighted.

4.5 Summary of Formulas

Ordered Weighted Averaging Operator (OWA)

$$OWA_{\mathbf{w}}(\mathbf{x}) = \sum_{i=1}^{n} w_i x_{(i)} = w_1 x_{(1)} + w_2 x_{(2)} + \cdots + w_n x_{(n)}$$

where $x_{(1)} \leq x_{(2)} \leq \cdots \leq x_{(n)}$.

Special Cases of the OWA

Minimum $\quad \mathbf{w} = \langle 1, 0, 0, \ldots, 0 \rangle$

Maximum $\quad \mathbf{w} = \langle 0, 0, \ldots, 0, 1 \rangle$

Median $\quad \mathbf{w} = \langle 0, \ldots, 0, 1, 0, \ldots, 0 \rangle, \quad n = 2k+1$
$\qquad\qquad\qquad\; \underbrace{}_{k \text{ zeros}} \quad\; \underbrace{}_{k \text{ zeros}}$

$\qquad\quad\;\; \mathbf{w} = \langle 0, \ldots, 0, 0.5, 0.5, 0, \ldots, 0 \rangle, \quad n = 2k$
$\qquad\qquad\quad\; \underbrace{}_{k-1 \text{ zeros}} \qquad\qquad \underbrace{}_{k-1 \text{ zeros}}$

Trimmed mean $\quad \mathbf{w} = \langle 0, \ldots, 0, \frac{1}{n-h}, \frac{1}{n-h}, \ldots, \frac{1}{n-h}, 0, \ldots, 0 \rangle$
$\qquad\qquad\qquad\;\; \underbrace{}_{\frac{h}{2} \text{ zeros}} \qquad\qquad\qquad\qquad\quad \underbrace{}_{\frac{h}{2} \text{ zeros}}$

Winsorized mean $\mathbf{w} = \langle 0, \ldots, 0, \frac{2+h}{2n}, \frac{1}{n}, \ldots, \frac{1}{n}, \frac{2+h}{2n}, 0, \ldots, 0 \rangle$
$\qquad\qquad\qquad\;\; \underbrace{}_{\frac{h}{2} \text{ zeros}} \qquad\qquad\qquad\qquad\qquad\quad \underbrace{}_{\frac{h}{2} \text{ zeros}}$

Orness of OWA

$$\text{orness}(\mathbf{w}) = \sum_{i=1}^{n} w_i \frac{i-1}{n-1}.$$

Choquet Integral

$$C_v(\mathbf{x}) = \sum_{i=1}^{n} x_{(i)} \left(v(\{(i) : (n)\}) - v(\{(i+1) : (n)\}) \right)$$

where $x_{(1)} \leq x_{(2)} \leq \cdots \leq x_{(n)}$.

Orness of Choquet Integral

$$\text{orness}(v) = \frac{1}{n-1} \sum_{S \subset \{1:n\}} \frac{(n-|S|)!|S|!}{n!} v(S).$$

Shapley Index

$$\phi(i) = \sum_{S \subseteq \{1:n\} \setminus \{i\}} \frac{(n-|S|-1)!|S|!}{n!} \left(v(S \cup \{i\}) - v(S) \right).$$

Then the Shapley value is

$$\phi = \langle \phi_1, \phi_2, \dots, \phi_n \rangle.$$

Special Cases of the Choquet Integral
 WAM if $v(S) = \sum_{i \in S} v(\{i\})$.
 OWA if $v(S) = v(T)$ whenever $|S| = |T|$.

4.6   Practice Questions

1. Calculate the outputs for an OWA operator with $\mathbf{w} = \langle 0.1, 0.4, 0.3, 0.2 \rangle$ and

   (i) $\mathbf{x} = \langle 0.8, 0.2, 1, 1 \rangle$
   (ii) $\mathbf{x} = \langle 0.9, 0.1, 0.7, 0.6 \rangle$

2. Calculate the outputs for an OWA operator with $\mathbf{w} = \langle 0.1, 0, 0.1, 0.3, 0.5 \rangle$ and

   (i) $\mathbf{x} = \langle 0.9, 0.3, 0.5, 0.8, 0 \rangle$
   (ii) $\mathbf{x} = \langle 0.2, 0.1, 0.1, 0.9, 1 \rangle$

3. Calculate the output for a discrete Choquet integral for $\mathbf{x} = \langle 0.8, 0.3, 0.1 \rangle$ when the associated fuzzy measure is given by

$$v(\{1, 2, 3\}) = 1$$

$$v(\{1, 2\}) = 0.9 \quad v(\{1, 3\}) = 0.2 \quad v(\{2, 3\}) = 0.3$$

$$v(\{1\}) = 0.2 \qquad v(\{2\}) = 0.3 \qquad v(\{3\}) = 0.1$$

$$v(\emptyset) = 0$$

4. Calculate the output for a discrete Choquet integral for $\mathbf{x} = \langle 1, 8, 12, 7 \rangle$ when the associated fuzzy measure is given by

$$v(\{1, 2, 3, 4\}) = 1$$

$$v(\{1, 2, 3\}) = 0.9 \quad v(\{1, 2, 4\}) = 0.9 \quad v(\{1, 3, 4\}) = 0.9 \quad v(\{2, 3, 4\}) = 0.9$$

$$v(\{1, 2\}) = 0.2 \quad v(\{1, 3\}) = 0.2 \quad v(\{1, 4\}) = 0.2$$

$$v(\{2, 3\}) = 0.9 \quad v(\{2, 4\}) = 0.2 \quad v(\{3, 4\}) = 0.3$$

$$v(\{1\}) = 0.2 \quad v(\{2\}) = 0.3 \quad v(\{3\}) = 0.1 \quad v(\{4\}) = 0.1$$

$$v(\emptyset) = 0$$

5. Define the weighting vectors for $n = 3$ and $n = 5$ using the quantifier $Q(t) = t^2$.
6. Compare the weighting vectors for $Q(t) = \sqrt{t}$ (Example 4.5) and $Q(t) = t^2$ in the previous question. For quantifiers of the form $Q(t) = t^q$, what can you predict about the orness of the weighting vectors when $q > 1$ or $q < 1$?
7. Define a weighting vector for $n = 7$ that can approximate the natural language quantifier "at least 3".
8. Define a weighting vector for $n = 4$ that can approximate the natural language quantifier "almost all".
9. The manager of a clothing store forecasts her monthly sales based on an OWA of the previous 4 months. Interpret the vectors $\mathbf{w} = \langle 0.1, 0.2, 0.3, 0.4 \rangle$ and $\mathbf{w} = \langle 0.4, 0.3, 0.2, 0.1 \rangle$ in terms of whether they give a 'conservative' or 'optimistic' estimate of the next month's sales.
10. What would be the fuzzy measure used to define a Choquet integral that is equivalent to an OWA function with weights $\mathbf{w} = \langle 0.2, 0.7, 0.1 \rangle$?
11. What would be the fuzzy measure used to define a Choquet integral equivalent to a WAM with $\mathbf{w} = \langle 0.3, 0.45, 0.25 \rangle$?
12. Will the fuzzy measure below define a Choquet integral that is equivalent to either the weighted arithmetic mean or the OWA? Explain why/why not.

$$v(\{1, 2, 3\}) = 1$$

$$v(\{1, 2\}) = 0.7 \quad v(\{1, 3\}) = 0.6 \quad v(\{2, 3\}) = 0.8$$

$$v(\{1\}) = 0.2 \quad v(\{2\}) = 0.5 \quad v(\{3\}) = 0.4$$

$$v(\emptyset) = 0$$

13. Will the fuzzy measure below define a Choquet integral that is equivalent to either the weighted arithmetic mean or the OWA? Explain why/why not.

$$v(\{1,2,3\}) = 1$$

$$v(\{1,2\}) = 0.7 \quad v(\{1,3\}) = 0.7 \quad v(\{2,3\}) = 0.7$$

$$v(\{1\}) = 0 \qquad v(\{2\}) = 0 \qquad v(\{3\}) = 0$$

$$v(\emptyset) = 0$$

14. Suppose we have four job applicants who want a job on our data analysis team. We need them to be *either* strong in both coding and mathematics, *or* strong communicators. They are given scores out of 10 for each of the criteria.

Candidate	JA 1	JA 2	JA 3	JA 4
Coding	9	3	2	7
Mathematics	2	8	3	8
Communication	4	6	8	3

(i) Define a fuzzy measure that can model our requirements. [Hint: the value assigned to *coding* and *mathematics* should be high but the values assigned to these two criteria individually should be very low. The value assigned to *communication* on its own should also be high.]

(ii) Evaluate the 4 candidates using the Choquet integral with respect to the fuzzy measure you defined in (i).

15. For the fuzzy measure below, which of the criteria seems most important overall? Which seems least important?

$$v(\{1,2,3\}) = 1$$

$$v(\{1,2\}) = 0.8 \quad v(\{1,3\}) = 0.9 \quad v(\{2,3\}) = 0.4$$

$$v(\{1\}) = 0.1 \qquad v(\{2\}) = 0.2 \qquad v(\{3\}) = 0.4$$

$$v(\emptyset) = 0$$

4.7 R Tutorial

The OWA operator and Choquet integral will be very useful for us to define functions in the R environment [13], since calculating them step by step can sometimes be very difficult. As with the weighted median from the last topic, we will need our sort(), order(), and rank() functions.

4.7.1   OWA Operator

Recall that we defined our weighted arithmetic mean using the following.

```
WAM <- function(x,w) {sum(w*x)}
```

The OWA function should be very similar, however first we need to re-order the vector **x** from lowest to highest. This is achieved using the sort() function, which reorders the vector non-decreasingly.[2]

```
OWA <- function(x,w) {
sum(w*sort(x))
}
```

We can make sure our weighting vector is normalized by including the line 'w <- w/sum(w)' at the beginning. This is also a situation where we may want to include a default value for the weighting vector. As with the weighted power mean, we could update the above to

```
OWA <- function(x,w=array(1/length(x),length(x)) ) {
   w <- w/sum(w)
   sum(w*sort(x))
}
```

The special cases of the OWA, the trimmed mean and Winsorized mean can also be defined. In fact, the trimmed mean can be implemented in R using the mean function with an additional argument trim. Recall that the trimmed mean was defined in terms of a parameter 'h', which was an even number determining how many values are removed from calculation. However sometimes it makes more sense to define the trimmed mean in terms of the percentage that is removed. For example, with 10 data, a value of trim=0.1 would remove the highest and lowest entry (10% from the top and 10% from the bottom) and for an input vector **x** we have

```
mean(x,trim=0.1)
```

We can calculate the Winsorized mean by replacing the upper and lower $h\%$ of entries in the vector which is equivalent to allocating higher weights to the upper and lower non-trimmed inputs. We first determine how many values to replace (we will assume h is given as a percentage here too and then find hn). We will require the floor function for this step in case hn does not result in an integer value. As a default, we set the h percentage to zero.

---

[2]To define the OWA in terms of a non-increasing permutation, i.e. if we want the first weight to be allocated to the highest input, we can use decreasing = TRUE as an optional argument when we sort, i.e. sort(x,decreasing = TRUE).

```
Wins.mean <- function(x,h=0) {      # 1. pre-defining the inputs
  n <- length(x)                    # 2. store the length of x
  repl <- floor(h*n)                # 3. how many data to replace
  x <- sort(x)                      # 4. sort x
  x[1:repl] <- x[repl+1]            # 5. replace lower values
  x[(n-repl+1):n] <- x[n-repl]      # 6. replace upper values
  mean(x)                           # 7. calculate the mean
}
```

With both of these functions, we could set them up so that we input the rmv and repl values instead of *h*. Either way is fine, as long as it is taken into account when using the functions to calculate outputs.

R Exercise 15 *Enter in the OWA, trimmed mean and Winsorized mean and verify that you get the same results for* **x** $=$ $\langle 0.3, 0.6, 0.8, 0.3, 0.4, 0.5 \rangle$ *when they are equivalent, e.g. for the trimmed mean and h* $=$ *0.17, the OWA weights should be* **w** $=$ $\langle 0, 0.25, 0.25, 0.25, 0.25, 0 \rangle$, *and for the Winsorized mean, the outputs should be the same when the OWA weights are* **w** $= \langle 0, 1/3, 1/6, 1/6, 1/3, 0 \rangle$.

4.7.2 Choquet Integral

The Choquet integral is another function that could be coded in a number of ways. Our aim is to have a function that takes an input vector **x**, a fuzzy measure v and then produces the output. However note here that the fuzzy measure v needs be entered as an argument, and so we will need to decide on how this is to be done.

The most straightforward way is to code v as a vector, and so we just need to decide on the ordering (because there is no linear order on subsets). The two most common orderings are either by cardinality, so that the order of the set values would be,

$$v(\emptyset), v(\{1\}), v(\{2\}), \ldots, v(\{n\}), v(\{1,2\}), v(\{1,3\}), \ldots, v(\{1,2,\ldots,n\}),$$

or another type of ordering is called 'binary ordering'. Binary ordering uses the form of a binary number to decide which elements are in the set at a given position. For example, the first 6 binary numbers are 000, 001, 010, 011, 100, and 101. If the digit on the far right is a 1, then the 1st element/variable is in the subset. If it is a zero, then 1 is not in the subset. If the digit second from the right is a 1, then the 2nd element/variable is in the set and so on. So corresponding with 000 is $v(\emptyset)$, corresponding with 001 is $v(\{1\})$, corresponding with 011 is $v(\{1,2\})$, and corresponding with 101 is $v(\{1,3\})$. This means the order of the values would be,

$$v(\emptyset), v(\{1\}), v(\{2\}), v(\{1,2\}), v(\{3\}), v(\{1,3\}), v(\{2,3\}), v(\{1,2,3\}), v(\{4\}), \ldots$$

Although this might seem more complicated, using this ordering actually makes it much easier for us to define our function, because the position of a set in the vector can be directly calculated and will not change when *n* changes.

The following makes use of the order() function.

```
Choquet <- function(x,v) { # 1. pre-defining the inputs
  n <- length(x)                  # 2. store the length of x
  w <- array(0,n)                 # 3. create an empty weight vector
  for(i in 1:(n-1)) {             # 4. define weights based on order
    v1 <- v[sum(2^(order(x)[i:n]-1))]     #
                                  # 4i. v1 is f-measure of set of all
                                  #     elements greater or equal to
                                  #     i-th smallest input.
    v2 <- v[sum(2^(order(x)[(i+1):n]-1))]  #
                                  # 4ii. v2 is same as v1 except
                                  #      without i-th smallest
    w[i] <-  v1 - v2                # 4iii. subtract to obtain w[i]
    }                           #
  w[n] <- 1- sum(w)             # 4iv. final weight leftover
  x <- sort(x)                  # 5. sort our vector
  sum(w*x)                      # 6. calculate as we would WAM
}
```

The steps for 4i to 4iii could be combined into a single step,

```
w[i] <- v[sum(2^(order(x)[i:n]-1))] - v[sum(2^(order(x)[(i+1):n]-1))]
```

Recall that the `order()` function tells us the indices in **x** corresponding with the lowest to highest inputs. So a result of $\langle 3, 1, 2, 4 \rangle$ would mean that x_3 is the lowest, x_1 is the second lowest, x_2 is the second highest and x_4 is the highest. Then taking `order(x)[3:4]` would give us the indices of the two highest inputs (2   4). Subtracting 1 and raising 2 to the power of this vector gives the corresponding position of our subset. So in the case of 2   4, subtracting 1 makes it 1   3 and 2 to the power of these values gives 2 and 8. Summing these together gives 10, which in binary is 1010, i.e. the 4th and 2nd element (don't forget that we read the binary number from right to left). So the fuzzy measure value associated with this subset should have been stored in the 10th position.

The last weight is just found by subtracting the previously defined weights from 1 rather than redefining v2 so that it doesn't try to take `order(x)[(n+1):n]` when what we actually want there is the empty set.

R Exercise 16 *Enter in the Choquet integral function (you can copy and paste if it is easier). Verify the examples used throughout the topic. In the Example from Sect. 4.4.4, the fuzzy measure shown would be encoded as*
`c(0.3, 0.4, 0.4, 0.1, 0.9, 0.5, 1)`.
We don't need to input the empty set (since we don't use it in calculation) and this allows the value of `v[1]` *to correspond with 001,* `v[2]` *to correspond with 010 and so on.*

4.7.3 Orness

Compared with the Choquet integral, the orness calculations for both the OWA and the Choquet integral should seem relatively easy. In the case of the Choquet integral, we will need to use the factorial function, which is calculated using `factorial()` (and not '!').

Firstly for the OWA, we multiply each weight by $\frac{i-1}{n-1}$. The vector of these values can be found just using our vector subtraction and division operations

```
orness <- function(w) {    # 1. the input is a weighting vector
n <- length(w)             # 2. store the length of w
sum(w*(1:n-1)/(n-1))       # 3. orness calculation
}
```

The orness calculation for the Choquet integral requires us to work out the cardinality of the corresponding subsets of the fuzzy measure. To do this, we will use a special function that converts any integer to its binary representation. This is intToBits(). The output of this function is a string of double digit values, e.g. 00   01   01   00 ... where a 01 corresponds with a 1 in the binary place value representation. This is done in the reverse order to how we normally write binary numbers. For example, 10011 would be the same as 01 01 00 00 01. We need to invoke the as.numeric() function on this vector in order to sum the values because, as a default, bits are not treated like numbers. This representation makes an operation like finding the cardinality based on the position of the subset very easy.

```
orness.v <- function(v) {    # 1. the input is a fuzzy measure
n <- log(length(v)+1,2)      # 2. calculates n based on |v|
m <- array(0,length(v))      # 3. empty array for multipliers
for(i in 1:(length(v)-1)) {  # 4. S is the cardinality of
  S <- sum(as.numeric(intToBits(i))) #    of the subset at v[i]
  m[i] <- factorial(n-S)*factorial(S)/factorial(n)   #
  }                          #
sum(v*m)/(n-1)               # 5. orness calculation
}
```

R Exercise 17 *Verify the following orness calculations (assuming the input is associated with an OWA function).*

Input (orness(...))	Expected output
c(1,0,0,0)	*0*
c(0,0,0,0,1)	*1*
c(0,0.2,0.6,0.2,0)	*0.5*
((1:4)/4)^2 - ((0:3)/4)^2	*0.7083333*

R Exercise 18 *Verify the following orness calculations (assuming the input is a fuzzy measure associated with a Choquet integral).*

Input (orness.v(...))	Expected output
c(0,0,0,0,0,0,1)	*1*
c(1,1,1,1,1,1,1)	*0*
c(0.3,0.1,0.4,0.6,0.9,0.7,1)	*0.5*
c(0.3,0.4,0.4,0.1,0.9,0.5,1)	*0.4333333*

4.7.4   Shapley Values

For determining the Shapley value from a fuzzy measure, we will once again use the `intToBits()` function in order to calculate the cardinality. We also use it to determine whether or not a variable is in the corresponding subset.

```
shapley <- function(v) {    # 1. the input is a fuzzy measure
   n <- log(length(v)+1,2)     # 2. calculates n based on |v|
   shap <- array(0,n)          # 3. empty array for Shapley values
   for(i in 1:n) {             # 4. Shapley index calculation
      shap[i] <- v[2^(i-1)]*factorial(n-1)/factorial(n) #
                               # 4i.   empty set term
      for(s in 1:length(v)) {    # 4ii. all other terms
         if(as.numeric(intToBits(s))[i] == 0) {  #
                               # 4iii.if i is not in set s
         S <- sum(as.numeric(intToBits(s)))    #
                               # 4iv. S is cardinality of s
         m <- (factorial(n-S-1)*factorial(S)/factorial(n))  #
                               # 4v. calculate multiplier
         vSi <- v[s+2^(i-1)]      # 4vi. f-measure of s and i
         vS <- v[s]               # 4vii. f-measure of s
         shap[i]<-shap[i]+m*(vSi-vS) # 4viii. add term
         }                       #
      }                        #
   }                         #
   shap                      # 5. return shapley indices
}                          #     vector as output
```

R Exercise 19 *Verify the following Shapley value calculations.*	
Input (`shapley(...)`)	*Expected output*
`c(0.4,0.1,1)`	*0.65 0.35*
`c(0.3,0.2,0.5,0.5,0.8,0.7,1)`	*0.3 0.2 0.5*
`c(0.3,0,0.5,0.2,0.3,1,1)`	*0.2 0.4 0.4*
`c(0.1,0.4,0.7,0.3,`	
`    0.5,0.8,0.9,0.1,0.1,`	*0.158333 0.475 0.30833 0.05833*
`    0.5,0.7,0.3,0.5,0.8,1)`	

4.8   Practice Questions Using R

1. Use `rand.x <- sample(1:1000,10)` to create a random input vector and calculate the OWA with weights defined by a quantifier using
   `w <- ((1:10)/10)^2 - ((0:9)/10)^2`.
2. Use the method from the previous question to compare the outputs of OWA functions and power means. For power means with power *p* and quantifier-based

OWAs that use $Q(t) = t^q$, look at whether or not the functions tend toward higher or lower inputs. The quantifier in the previous question is based on $Q(t) = t^2$ (for $n = 10$), changing the 2 to a 3 will mean it is uses $Q(t) = t^3$ and so on).

3. Open the data file[3] FMEASURES.txt and assign the two columns as vectors to fm.1 and fm.2 (or any assignment you like—these are fuzzy measures for 5 variables).

   (i) Calculate the output for $\mathbf{x} = \langle 5, 15, 13, 8, 9 \rangle$ using fm.1 and fm.2.
   (ii) Calculate the Shapley values.
   (iii) Calculate the orness values.
   (iv) Explain why fm.1 or fm.2 resulted in a higher output with reference to the Shapley and orness calculations.

References

1. Beliakov, G., Pradera, A., Calvo, T.: Aggregation Functions: A Guide for Practitioners. Springer, Heidelberg (2007)
2. Beliakov, G., Bustince, H., Calvo, T.: A Practical Guide to Averaging Functions. Springer, Berlin/New York (2015)
3. Choquet, G.: Theory of capacities. Ann. Inst. Fourier **5**, 131–295 (1953)
4. Dujmovic, J.J.: Weighted conjunctive and disjunctive means and their application in system evaluation. Univ. Beograd. Publ. Elektrotechn. Fak. **461/497**, 147–158 (1974)
5. Gagolewski, M.: Data Fusion. Theory, Methods and Applications. Institute of Computer Science, Polish Academy of Sciences, Warsaw (2015)
6. Grabisch, M.: The applications of fuzzy integrals in multicriteria decision making. Eur. J. Oper. Res. **89**, 445–456 (1996)
7. Grabisch, M., Marichal, J.-L., Mesiar, R., Pap, E.: Aggregation Functions. Cambridge University Press, Cambridge (2009)
8. Hampel, F.R., Ronchetti, E.M., Rousseeuw, P.J., Stahel, W.A.: Robust Statistics: The Approach Based on Influence Functions. Wiley, Hoboken, NJ (2005)
9. Huber, P.J., Ronchetti, E.M.: Robust statistics, 2nd edn. Wiley, Hoboken, NJ (2009)
10. IMDb: Box Office Mojo (movie statistics) http://www.boxofficemojo.com/movies/ (2016). Cited 17 April 2016
11. Mesiar, R., Stupňanová, A.: Integral sums and integrals. In: Torra, V., Narukawa, Y., Sugeno, M. (eds.) Non-additive Measures. Studies in Fuzziness and Soft Computing, vol. 310, pp. 63–78. Springer, Switzerland (2014)
12. Panaligan, R., Chen, A.: Quantifying Movie Magic with Google Search. Google Whitepaper: Industry Perspectives and User Insights. Available online http://ssl.gstatic.com/think/docs/quantifying-movie-magic_research-studies.pdf (2013). Cited 31 July 2016
13. R Core Team: R: A language and environment for statistical computing. R Foundation for Statistical Computing, Vienna. http://www.R-project.org/ (2014)

---

[3]Can be found at http://www.researchgate.net/publication/306099814_AggWAfit_R_library or alternatively, http://aggregationfunctions.wordpress.com/book. These can be saved to your R working directory.

14. Torra, V., Narukawa, Y.: Modeling Decisions. Information Fusion and Aggregation Operators. Springer, Berlin/Heidelberg (2007)
15. Walden, P.: Moviepilot Digital Tracking. Variety (online magazine) http://variety.com/t/moviepilot/ (2016). Cited 17 April 2016
16. Yager, R.R.: On ordered weighted averaging aggregation operators in multicriteria decision making. IEEE Trans. Syst. Man Cybern. **18**, 183–190 (1988)

Chapter 5
Fitting Aggregation Functions to Empirical Data

When aggregation functions are used to allocate scores, e.g. when a weighted arithmetic mean is used to summarize a university student's achievement across each of his or her subjects, it is sometimes straightforward to decide on an appropriate weighting vector and an appropriate class or family of functions. In the first chapter, we saw situations that necessitated geometric or harmonic means, and we also introduced the notion of quantifier guided aggregation in Chap. 4. However there are situations where we might not know appropriate weights, or we may have collected data and we want to know the best choice of weights in order to model the relationship between inputs and output.

The theory of aggregation functions can hence be useful beyond data summarization and scoring systems, it can also be applied in order to analyze data, discover hidden relationships and to make predictions based on previous patterns. In statistics, the key tool or method for these tasks is known as regression, while in broader computer science research, these goals are essentially those of machine learning and data mining.

In this chapter we will focus on the use of optimisation in order to learn the ideal weights associated with a given aggregation function in order to model a dataset. The weights we learn can then be used either to predict the output for a new observation, for example, in the previous topic we looked at predicting the box office success for movies based on online analytics and other data, or we can also interpret the weights toward the importance and interaction of each variable in the learned model. The key here is to be reasonable when it comes to interpreting the weights we find, avoiding conclusions that overstate the significance of our results.

Assumed Background Concepts

- Distance
  How do you calculate the length for the hypotenuse of a triangle or the diagonal of a prism?

© Springer International Publishing AG 2016
S. James, *An Introduction to Data Analysis using Aggregation Functions in R*,
DOI 10.1007/978-3-319-46762-7_5

> **Chapter Objectives**
>
> - Be able to apply data fitting code in order to define parameters and interpret datasets
> - To be able to make reasonable assessments of goodness of fit and accuracy of models

5.1   The Problem in Data: Recommender Systems

Collaborative-based recommender systems used by online sites (e.g. Amazon.com [8]) use ratings and purchase history of 'similar' users to predict how you will respond to an unseen item and provide personalized recommendations for new products.

Suppose Kei is a member of a hotel booking website. She has done a search for "Melbourne" and there are three potential hotels that have all been rated by other members who have rated a number of hotels in common with Kei. The recommender system wants to be able to say something like "based on similar users, we believe you will enjoy staying at Hotel X".

	Kei's	Similar user ID								
Hotel ID	reviews	220	817	751	265	656	231	289	345	171
159	56	65	18	56	69	58	53	61	70	50
508	73	41	31	61	78	78	75	83	73	78
457	81	60	100	71	69	66	91	74	100	90
215	83	73	84	90	90	86	81	54	96	89
343	56	80	76	40	43	49	62	38	52	86
299	79	83	76	59	67	75	80	79	87	35
277	92	67	58	80	90	95	93	100	100	100
242	99	69	99	91	100	84	100	100	92	96
2		98	50	62	63	81	72	82	96	51
826		94	99	63	58	96	70	91	59	86
977		59	85	66	55	78	91	87	94	73

To make this recommendation, the system needs to predict the score for these unseen hotels based on the similar users. We can use aggregation functions to provide these scores, however we probably want to account for the fact that some users will be more similar than others, i.e. we might wish to incorporate weights.

There are a number of ways to choose weights. When the data set gets very large, making judgements as we did previously (e.g. for student assessment and to account for manager importance) becomes impractical. One popular automated method in recommender systems is to use a 'similarity' calculation based on the notion of distance.

> **Definition (informal) 5.1 (Distance).** A distance function defined for two arguments (or points) gives us an idea of how 'far apart' two objects are. If the two arguments are mathematically identical, then the distance should be zero, and as the objects are moved apart the distance should be positive. Distances are symmetric (so the distance from A to B should be the same as from B to A), and also satisfy a property referred to as the **triangular inequality**, which requires that the distance calculated between two points directly should not be greater than the distance when calculated via a third point.

> **Definition 5.1 (Distance).** For two inputs $\mathbf{x} = \langle x_1, x_2, \ldots, x_n \rangle, \mathbf{y} = \langle y_1, y_2, \ldots, y_n \rangle$, a function $d(\mathbf{x}, \mathbf{y})$ is a distance if:
>
> - $d(\mathbf{x}, \mathbf{y}) \geq 0$ for all inputs and $d(\mathbf{x}, \mathbf{y}) = 0$ if and only if $\mathbf{x} = \mathbf{y}$;
> - $d(\mathbf{x}, \mathbf{y}) = d(\mathbf{y}, \mathbf{x})$;
> - with respect to a third vector $\mathbf{z} = \langle z_1, z_2, \ldots, z_n \rangle$, it holds that
>
> $$d(\mathbf{x}, \mathbf{y}) \leq d(\mathbf{x}, \mathbf{z}) + d(\mathbf{y}, \mathbf{z}).$$

For real numbers, the distance is usually just given by the difference (e.g. the distance between 3 and 9 is 6), however for vectors there are a number of distances we could use. A useful paramaterized function is the **Minkowski distance**. For a given $p \geq 1$, the distance between Kei and the j-th user in terms of their ratings can be denoted $d_p(\text{kei}, j)$ and given by,

$$d_p(\text{kei}, j) = \left(\sum_{i=1}^{n} |x_{i,\text{kei}} - x_{i,j}|^p \right)^{1/p},$$

where $x_{i,\text{kei}}$ are $x_{i,j}$ are Kei and the j-th users' ratings for the i-th hotel.

> **Notation Note** Distance on multidimensional inputs
> Implied by the form of the Minkowski distance is that the vectors need to be of the same dimension. So we can't measure the distance between, say, a value 4 and a vector $\langle 2, 5 \rangle$. The value of p needs to be larger or equal to 1 in order for the triangular inequality to be satisfied. Remember that the output of a distance function should be a value, even if the inputs are vectors.
> The $|\cdot|$ brackets here were used in the previous chapter to give the size or cardinality of a set. In this case however, it means that if the evaluation inside produces a negative number, then we make it positive (or leave it as a positive number if it is above zero). For example, $|2 - 6| = |-4| = 4$, which is the same as $|6 - 2| = |4| = 4$.

You might recognise this as similar to the power mean of the differences between each score (without dividing through by n). For example, the distance between Kei and user 220 when $p = 2$ would be calculated as

$$\left(|56 - 65|^2 + |73 - 41|^2 + |81 - 60|^2 + \ldots + |99 - 69|^2 \right)^{1/2}.$$

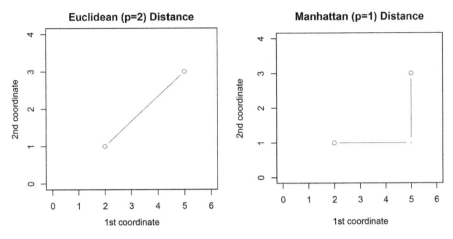

Fig. 5.1 Depictions of the Euclidean and Manhattan distances for 2-dimensional points. In the case of Euclidean, we calculate the direct path ($\sqrt{3^2 + 2^2} \approx 3.606$), while for the Manhattan we can only travel along the direction of the co-ordinate axes ($3 + 2 = 5$)

When $p = 2$, the Minkowski distance becomes the Euclidean distance, while $p = 1$ corresponds with the special case of the Manhattan or taxi-cab distance (Fig. 5.1).

> **R Exercise 20** *Write a function in* R *[10] for evaluating the Minkowski distance between two vectors and verify the outputs in the figure. [Hint: You can use* abs() *for determining the absolute value, e.g.* |**x** − **y**| *can be calculated using* abs(x - y) *where* **x** *and* **y** *are vectors. Remember that you can set a default value of p using, e.g.*
> ```
> minkowski <- function(x,y,p=1) {...}
> ```

We can then define the weights of the similar users relative to these distances. For example, if the distances to three similar users were $\langle 24, 39, 81 \rangle$, then we could transform them by taking a negation or decreasing function of each value (so that less distance or a 'more similar' user is allocated a higher weight) and then scale them so that they add to 1. For example, using a negation of the form we used in Chap. 2, $f(t) = 81 - t + 24$, we would obtain $\langle 81, 66, 24 \rangle$ which becomes $\langle 0.47, 0.39, 0.14 \rangle$ when we divide through by the sum. We could also take the reciprocal, e.g. $\langle \frac{1}{24}, \frac{1}{39}, \frac{1}{81} \rangle$, before scaling.

> **Side Note 5.1** *Another popular measure of similarity is known as cosine similarity (from vector arithmetic), which is given by,*

$$\cos(\text{kei}, j) = \frac{\sum\limits_{i=1}^{n} x_{i,kei} \cdot x_{i,j}}{\left(\left(\sum\limits_{i=1}^{n} x_{i,kei}^2\right) \cdot \left(\sum\limits_{i=1}^{n} x_{i,j}^2\right)\right)^{1/2}}.$$

Two vectors are considered to be in the same 'direction' and will have a cosine similarity equal to 1 if their components are in the same proportion. For example, $\langle 1, 2, 4 \rangle$ is in the same direction as $\langle 3, 6, 12 \rangle$, however it is in the opposite *direction to $\langle -2, -4, -8 \rangle$, so the cosine similarity of these would be -1.*

Whilst calculations based on distance (or some other form of similarity) between users can be effective, sometimes we can find weights that perform better in terms of predicting accurate scores. These ideal weights may not necessarily correspond with our similarity calculations. For example, similar users to *Kei* might also be similar to each other and their scores may become redundant in terms of providing useful information upon which to base our predictions. A better combination of weights may be one that takes a broader spectrum of users into account.

When we have a dataset consisting of observed vector inputs (e.g. similar users' ratings) and corresponding outputs (Kei's scores), we can use parameter identification or optimized 'fitting' techniques that find the best combination of weights. For Kei's hotel ratings data, using such a method we obtain the weighting vector

$$\mathbf{w}^f = \langle 0.035, 0.049, 0, 0.033, 0.347, 0.389, 0.050, 0.097, 0 \rangle.$$

We might then interpret this weighting vector as suggesting that the ratings of users 656 and 231 (corresponding with weights w_5 and w_6) are the most influential in predicting Kei's ratings. This does not *necessarily* mean that they are the most similar—rather that the combination of their scores is the most useful in our model.

If we instead found the weights using the Manhattan distance (Minkowski distance with $p = 1$), subtracting each of the totals from 183 (since the maximum distance is 157 and the minimum is 26) and then normalizing these values by dividing through by the sum, we obtain,

$$\mathbf{w}^d = \langle 0.033, 0.031, 0.116, 0.140, 0.153, 0.186, 0.125, 0.130, 0.085 \rangle.$$

From these we see that according to this distance measure, the same two users are the most similar, however we also note that even though the third similar user (user 751) is 'closer' to Kei than the first (user 220), the weight in the fitted weighting vector was higher for the first user. Let's now compare the predicted scores using these weighting vectors.

Kei's ratings	Predicted ratings using weights obtained through optimization ($\mathbf{w}^f$)	Predicted ratings using weights calculated from Manhattan distances ($\mathbf{w}^d$)
56	56	59.63
73	73	74.81
81	81	79.95
83	83	83.38
56	56	52.00
79	78.37	72.11
92	92	93.38
99	92.56	94.62

Except for the last hotel, the weighting vector obtained using the fitting algorithm is better at predicting Kei's rating (in 6 out of 8 cases they are obtained exactly). However we need to also keep in mind that we were finding 9 weights for only 8 instances—there is a high chance that even if our fitted weights and distance-based weights provide good accuracy here, they could be highly biased towards this data set. This problem is known as *overfitting*.

In this chapter we will learn about fitting methods such as the one used here to find $\mathbf{w}^f$. We will also discuss some considerations we should have in mind in interpreting our results. Fitting aggregation functions to data is just one method that can be used either for regression-type problems (learning the weights in order to predict the ratings) or classification problems (when we predict a categorical variable or class rather than a numerical output). Our emphasis is on being able to make reasonable judgements of assessments that may either be useful in themselves or provide a good starting point before using more complicated and involved methods. Whereas in previous chapters we included a separate R tutorial, here we will be making direct reference to R commands and packages (as well as including short exercises) throughout.

5.2 Background Concepts

We will make use of some linear programming optimization techniques [13] in order to fit aggregation functions to data. We will give a brief overview of how these linear optimization tasks can be set up and then discuss how to interpret and use the results for analyses.

5.2.1 Optimization and Linear Constraints

A linear program consists of an *objective equation* that is to be minimized (or maximized) and a set of inequalities. The objective is of the form:

$$z_1 v_1 + z_2 v_2 + \cdots + z_n v_n$$

where v_j are the decision variables, i.e. the unknown parameters that we want to find the optimum values for, and z_j are constants (depending on the context) affecting how much the objective is increased or decreased with changes in v_j. The restrictions on the range of choices for **v** are modelled as inequality constraints:

$$a_{i,1} v_1 + a_{i,2} v_2 + \cdots + a_{i,n} v_n \geq c_i,$$

or in some cases we may have '=' or '$\leq$' instead of '$\geq$'.

Definition 5.2 (Linear Program). For a given vector of decision variables $\mathbf{v} = \langle v_1, v_2, \ldots, v_n \rangle$, fixed matrices $\mathbf{A}, \mathbf{B}$ and constant vectors $\mathbf{c}, \mathbf{d}, \mathbf{z}$, a linear program (LP) is given by

$$\text{Minimize } \mathbf{v} \cdot \mathbf{z}$$

$$\text{s.t. } \mathbf{Av} = \mathbf{c}$$

$$\mathbf{Bv} \geq \mathbf{d}$$

Notation Note Matrix multiplied by a vector
We have not previously discussed how multiplication between a matrix $\mathbf{A}$ and $\mathbf{v}$ should be defined. In expanding out $\mathbf{Av} = \mathbf{c}$, we will obtain the set of equations

$$a_{1,1} v_1 + a_{1,2} v_2 + \cdots + a_{1,n} v_n = c_1$$
$$a_{2,1} v_1 + a_{2,2} v_2 + \cdots + a_{2,n} v_n = c_2$$
$$\cdots$$
$$a_{m,1} v_1 + a_{m,2} v_2 + \cdots + a_{m,n} v_n = c_m.$$

The objective equation and constraints array are hence all in this linear form.

To solve linear programs using software, we provide the objective equation along with the inequality and equality constraints (usually these are provided simply as the vectors $\mathbf{z}, \mathbf{c}, \mathbf{d}$ and the matrices $\mathbf{A}, \mathbf{B}$). The solver then returns the best choice of $\mathbf{v}$ that satisfies the constraints and gives the optimum value for our objective equation.

R Exercise 21 *In R we can use the lpsolve package [3], which includes the* `lp()` *function. The first argument is either* `"min"` *or* `"max"`, *then the objective coefficients* $\mathbf{z}$, *the entire constraints matrix (**A** and **B** combined), a*

vector of equalities or inequalities (e.g. c("`>=`","`==`","`==`","`<=`")*) and
the right hand side of the inequalities (***c** and **d** combined).
Load the lpsolve package and then enter in (on one line)*
```
lp("min",c(0.4,0.8),rbind(c(1,1),c(0.3,0)),
     c("==","<="),c(1,0.25))$solution
```
This will solve the LP:

$$\text{Minimize } 0.4v_1 + 0.8v_2$$

$$\text{s.t. } v_1 + v_2 = 1$$

$$0.3v_1 \ \leq 0.25$$

*Try changing some of the inputs to solve alternative LPs. The number
of rows of the constraints matrix should be the same as the length of the
inequalities vector and right-hand side values vector. The number of columns
should be the same as the length of the objective coefficients vector and the
number of decision variables.*

When we are fitting weighting vectors for aggregation functions, the values of **w**
will be our main decision variables. For choosing our objective function, common
approaches aim to minimize the error between predicted and observed values,
however 'error' itself can be quantified in a number of ways. The most common
choices are the squared differences or squared residuals.

We assume that we have the data in a format similar to our hotel ratings table,
with observed values for each variable (the hotel ratings given by other users) and
observed 'outputs' (Kei's ratings) that we want our function to approximate. More
generally, our data can be set out as follows, where $x_{i,j}$ represents the observed value
for the i-th instance and the j-th variable, and y_i is the output for that instance.

Values for each variable						Output
$x_{1,1}$	$x_{1,2}$	$x_{1,3}$	$\ldots$	$x_{1,n-1}$	$x_{1,n}$	y_1
$x_{2,1}$	$x_{2,2}$	$x_{2,3}$	$\ldots$	$x_{2,n-1}$	$x_{2,n}$	y_2
$x_{3,1}$	$x_{3,2}$	$x_{3,3}$	$\ldots$	$x_{3,n-1}$	$x_{3,n}$	y_3
$\ldots$						$\ldots$
$x_{m,1}$	$x_{m,2}$	$x_{m,3}$	$\ldots$	$x_{m,n-1}$	$x_{m,n}$	y_m

Our aim then is to find a weighted aggregation function $A_\mathbf{w}$ such that $A_\mathbf{w}(\mathbf{x}_i)$ (our
'predicted' value for the i-th observation) is as close to each of the y_i as possible
(minimizing the total error) for all of the instances. If we use squared differences as
our measure of 'closeness', we will have the following objective:

$$\text{Minimize } \sum_{i=1}^{m}(A_\mathbf{w}(\mathbf{x}_i) - y_i)^2.$$

However this objective is not linear with respect to our decision variables. If we consider a weighted arithmetic mean of two variables, the error for each instance becomes

$$(w_1 x_{i,1} + w_2 x_{i,2} - y_i)^2 = w_1^2 x_{i,1}^2 + w_2^2 x_{i,2}^2 + y_i^2 + 2w_1 w_2 x_{i,1} x_{i,2} - 2w_1 x_{i,1} y_i - 2w_2 x_{i,2} y_i.$$

We note that here we have w_1^2, w_2^2 and $w_1 w_2$ all present, which is what makes this expression non-linear.

> **Side Note 5.2** *Note that having $x_{i,1}^2$, y_i^2 etc do not pose a problem since these will all be fixed values based on our data (not variables).*

This kind of objective actually results in a **quadratic program** (QP) (provided the constraints are all linear). There are a number of standard approaches and software for solving them (e.g. in R we can use the quadprog package [12]), however we will opt for a linear approach. Instead of the squared differences, we sum the absolute values $|A_\mathbf{w}(x_i) - y_i|$, so our objective becomes:

$$\text{Minimize} \sum_{i=1}^{m} |A_\mathbf{w}(\mathbf{x}_i) - y_i|.$$

This objective is referred to as the **least absolute deviation** (LAD) or **least absolute error**. A special trick is then used to express this objective so that it can be solved by optimization software. It involves introducing two decision variables for each instance in our data set, r_i^+ and r_i^-, which will express either the positive or negative value resulting from $A_\mathbf{w}(x_i) - y_i$ (i.e. the residual). For each instance we then add the constraint:

$$A_\mathbf{w}(x_i) - r_i^+ + r_i^- = y_i.$$

Instead of our objective being in terms of the w_j, it will simply be the sum of all the r_i^+, r_i^- decision variables. However the values of w_j will still influence the objective since they influence the r_i^+, r_i^- via the above constraints, and we still include our constraint that ensures the w_j add to 1 (more detail about this approach can be found in [2]).

We can summarize the form of the linear program as follows for a dataset with m observations and n variables:

$$\text{Minimize} \quad \sum_{i=1}^{m} r_i^+ + r_i^-$$

$$\text{s.t.} \quad \left(\sum_{j=1}^{n} w_j x_{i,j} \right) - r_i^+ + r_i^- = y_i, \qquad i = 1, 2, \ldots, m$$

$$\sum_{j=1}^{n} w_j = 1$$

$$w_j \geq 0, \qquad j = 1, 2, \ldots, n$$

$$r_i^+, r_i^- \geq 0$$

This program takes the dataset and finds the weights of a weighted arithmetic mean that can predict the y_i in a way that minimizes the absolute error.

Notation Note Indexing with j

We use j as our index for the weights here (and throughout the chapter) because we have set our data table out such that each row represents an instance and each column represents a variable. So w_1 is the weight that applies to the first variable or column of data, w_2 to the second and so on.

In our opening example, we used this method to find the weighting vector for a weighted arithmetic mean such that WAM$(\mathbf{x}_i)$ was the same as the observed value y_i almost all of the time (except when Kei's rating was $y_i = 99$ and the predicted value was WAM$(\mathbf{x}_i) = 92.56$). Let's do another short example to help get an understanding of how the linear program works and how we can identify the vectors and matrices that need to be provided to functions like lp().

Example 5.1. Suppose we have the following data.

Observed inputs			Output
3	4	3	5
2	4	5	6
3	5	8	4
2	6	3	8

What will be the best choice of weights for predicting the outputs using a weighted arithmetic mean?

Solution. In this case the data has three variables (as inputs) so we will need decision variables for weights w_1, w_2 and w_3. We have 4 observed data so we will need 4 pairs of decision variables corresponding with the residuals, $r_1^+, r_1^-, \ldots, r_4^+, r_4^-$.

The first constraint that models WAM$(\mathbf{x}_1) - r_1^+ + r_1^- = y_1$ would explicitly be expressed

$$3w_1 + 4w_2 + 3w_3 - r_1^+ + r_1^- = 5.$$

In addition to constraints for each of the other data, we also need our constraint ensuring the weights add to 1,

$$w_1 + w_2 + w_3 = 1$$

To supply this information to `lp()`, we provide the information as vectors and matrices of coefficients. If a variable isn't included in a constraint or the objective, we write its coefficient as 0. The objective function for this data is hence supplied as the vector

$$\langle 0,0,0,1,1,1,1,1,1,1,1 \rangle.$$

The three 0s correspond to our weights, and the 1s correspond to the 4 pairs of residuals r_1^+, r_1^-. The constraints array that we provide will be

$$\begin{bmatrix} 3 & 4 & 3 & -1 & 1 & 0 & 0 & 0 & 0 & 0 & 0 \\ 2 & 4 & 5 & 0 & 0 & -1 & 1 & 0 & 0 & 0 & 0 \\ 3 & 5 & 8 & 0 & 0 & 0 & 0 & -1 & 1 & 0 & 0 \\ 2 & 6 & 3 & 0 & 0 & 0 & 0 & 0 & 0 & -1 & 1 \\ 1 & 1 & 1 & 0 & 0 & 0 & 0 & 0 & 0 & 0 & 0 \end{bmatrix},$$

the inequalities vector will be $\langle =,=,=,=,= \rangle$, and the right hand side vector will be $\langle 5,6,4,8,1 \rangle$ (the 4 y_i values and then 1 for the right hand side of the equation ensuring the weights sum to 1).

Using `lp()`, the solution output is

$$0\ 1\ 0\ 0\ 1\ 0\ 2\ 1\ 0\ 0\ 2$$

which indicates that the best weighting vector is $\langle 0,1,0 \rangle$, i.e. giving all weight to the second variable. This results in a total error of $1+2+1+2 = 6$. Note that the residuals for each instance always give the difference between $x_{i,2}$ and y_i (since the resulting WAM just bases its prediction on the second variable), e.g. for the first instance the value for the second input is 4, the output y_i is 5, and the residual values are hence $r_1^+ = 0, r_1^- = 1$.

5.2.2 Evaluating and Interpreting a Model's Accuracy

After obtaining the optimized values of $\mathbf{w}$, we might want to know how good the 'fit' is and how much overall error there is. This can be done in terms of the minimization criteria, e.g. the total error was 6 in the previous example, or can be with respect to some other calculation. For a fitted aggregation function $A_{\mathbf{w}}(\mathbf{x})$ and m instances, common error and goodness-of-fit measures include:

Sum of Squared Errors (SSE)   All of the differences between the predicted and observed output values, squared and then added together,

$$\text{SSE} = \sum_{i=1}^{m} (A_{\mathbf{w}}(\mathbf{x}_i) - y_i)^2.$$

Root Mean Squared Error (RMSE)   After calculating the Total squared error, we can divide by the number of observations (m) and then take the square root,

$$\text{RMSE} = \sqrt{\sum_{i=1}^{m} \frac{(A_{\mathbf{w}}(\mathbf{x}_i) - y_i)^2}{n}}.$$

This gives a value that can be interpreted as the *average difference* between each prediction and the output (it is actually the quadratic mean, a power mean with $p = 2$, so it will be affected more by larger differences). If a fitted function performs better than another in terms of RMSE, then it will also have a lower total squared error.

Sum of Absolute Errors (SAE)   The sum of all the absolute differences between predicted and observed outputs,

$$\text{SAE} = \sum_{i=1}^{m} |A_{\mathbf{w}}(\mathbf{x}_i) - y_i|.$$

Average Absolute Error (Av. AE)   This is the SAE divided by the number of observations,

$$\text{Av.AE} = \sum_{i=1}^{m} \frac{|A_{\mathbf{w}}(\mathbf{x}_i) - y_i|}{n}.$$

As with the RMSE, it can be interpreted as the average difference between each prediction and the output. Comparative goodness-of-fit will be the same for functions whether they are assessed using average absolute error or SAE.

Pearson Correlation (r)   This value between -1 and 1, gives an idea of how close the relationship between the two variables is to a linear relationship [9]. A perfect positive relationship will have $r = 1$, a perfect negative relationship will have $r = -1$ and no relationship will correspond with $r = 0$. The correlation between the vector of observed y (a) and vector of predictions $A_{\mathbf{w}}(\mathbf{x})$ (b) can be calculated in R using `cor(a,b,method="pearson")`, or by implementing the formula:

$$r = \frac{\sum_{i=1}^{m}(A_{\mathbf{w}}(\mathbf{x}_i) - \bar{A})(y_i - \bar{y})}{\sqrt{\sum_{i=1}^{m}(A_{\mathbf{w}}(\mathbf{x}_i) - \bar{A})^2}\sqrt{\sum_{i=1}^{m}(y_i - \bar{y})^2}}$$

where $\bar{A}$ is the arithmetic mean of the predicted outputs and $\bar{y}$ is the arithmetic mean of the observed y_i values.

When using Pearson correlation as a goodness-of-fit measure for aggregation functions, we would not be expecting to find a negative relationship, and in general we want the predicted and observed outputs to be as close as possible.

Spearman Correlation (ρ) Similar to Pearson's correlation, the Spearman correlation coefficient (ρ, pronounced 'rho') gives an indication of whether there is a monotone relationship between the observed and predicted outputs [11]. We may not have a linear relationship such that increases in the observed outputs correspond with equal increases to our predicted outputs, however it may be the case that if one observation y_1 is higher than another y_2, then we might want it to hold that our predicted output for $A_{\mathbf{w}}(\mathbf{x}_1)$ is higher than that predicted for $A_{\mathbf{w}}(\mathbf{x}_2)$. It can be calculated in R using cor(a,b,method="pearson") or with the same formula as Pearson's correlation, however with the relative rankings of the observed and predicted outputs instead of the values themselves. For example, if our observed outputs were $\langle 0.5, 0.95, 0.3, 0.72 \rangle$ and the predicted outputs were $\langle 0.7, 0.89, 0.66, 0.72 \rangle$, the Spearman's rank correlation is $\rho = 1$ since the relative orderings in both cases are $y_3 \prec y_1 \prec y_4 \prec y_2$.

R Exercise 22 *Calculate the goodness-of-fit using each of the above for*

$$\mathbf{a} = \langle 90, 40, 69, 31, 39, 44, 21, 81, 25, 52 \rangle$$

$$\mathbf{b} = \langle 64, 42, 65, 4, 55, 42, 18, 79, 46, 62 \rangle$$

and verify that you obtain the expected output.

Goodness-of-fit calculation	Implementation	Expected result
SSE	sum((a-b)^2)	2239
RMSE	sqrt(sum((a-b)^2)/length(a))	14.96329
SAE	sum(abs(a-b))	113
Av.AE	sum(abs(a-b))/length(a)	11.3
r	cor(a,b)	0.7711654
ρ	cor(a,b,method="spearman")	0.8024353

We used cor(a,b) *instead of* cor(a,b,method = "pearson") *here since* cor() *calculates Pearson's correlation as a default.*

The evaluation measure we choose to use may vary depending on our application. If we plan to use the aggregation model we constructed to rank different alternatives, then we may be more interested in using Spearman correlation, while if we want to create an aggregation function that predicts the average temperature in a room, we may be more interested in RMSE or Av.AE.

5.2.3   Flexibility and Overfitting

The more 'flexible' a function is, the easier it is for an optimization approach to find a function that fits the data well. This might be acceptable if we are only building an aggregation model to interpret the importance of our variables, however if we want to use our fitted aggregation function to predict new values, flexibility comes with the drawback of potential 'overfitting'.

Overfitting occurs when our model is too biased toward the data we have provided and not robust enough to reliably predict new values. For example, suppose we have data points with one input and one output, given by $(0, 0)$, $(1, 1)$, $(2, 4)$ and $(3, 9)$ and we want to find the best parameter β such that $y = \beta x$. If we only knew about the first two values when defining our model, we might find that $\beta = 1$ and the function $y = x$ would be perfect. However then when we use it to predict the outputs for $x = 2$ and $x = 3$ it will no longer be reliable.

We can see that there are actually two issues, which occurring simultaneously can lead to this problem: (1) not enough empirical data; and (2) the model having too many parameters and flexibility such that it *can* fit the data well, but might not generalize well to new data. We can illustrate this problem with the graphs below, which both show models for data with an approximate relationship $y = x$ (Fig. 5.2).

In order to determine whether our model is likely to overfit the data, we can make a reasonable assessment based on the ratio of observations to the number of parameters in the model. The number of observations should be at least as high and preferably *much* higher.

We can take a number of steps to avoid overfitting if it is seen to be a problem for what we are trying to achieve (i.e. if we are hoping to use our new model to predict unknown values). One option involves splitting our observed data into 'train' and

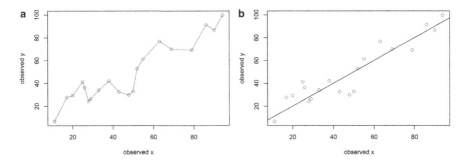

Fig. 5.2 The *line* in (**a**) represents a flexible model that is able to interpolate the data, however we can imagine that any new values are very unlikely to lie on this line. In fact, in cases where the original error in the data was high, the prediction for new values is likely to be even more inaccurate. On the other hand, in (**b**), the *line* represents a model which does not fit each of the data points individually, but would provide a predication that shouldn't be "too far" from any new values

'test' data. We use the training data to build our model and then the test data to evaluate how accurate it is. The best performing model will hence be one that is good at predicting unseen values. The split may be an important consideration, i.e. we could use 50 % test and 50 % training data, or we could use 10 % test and 90 % training. A popular choice in research is to use a 20–80 % split. Evaluation of a model can involve multiple iterations of randomly splitting the data and testing its performance. In doing so, however, bear in mind that the inferred focus becomes more about how reliable the general model or framework is rather than the specific models that we learn.

When the amount of training data is small, a weighted arithmetic mean may be more reliable than the Choquet integral for predicting unseen data because the Choquet integral is more flexible and hence will be liable to overfitting. However this does not mean that the Choquet integral is less appropriate when it comes to modelling the data, it just means we need a good amount of data in order to learn the best parameters.

5.3 Learning Weights for Aggregation Functions

Here we will specifically look at learning the weighting vectors for our power means and OWA operators, as well as fuzzy measures for the Choquet integral. We assume that we have access to a multivariate data set with observed instances $x_{i,1}, x_{i,2}, \ldots, x_{i,n}, y_i$, and that the data has been appropriately transformed (if necessary) so that the inputs and output are all given over the same scale. In general, we simply adapt the linear programming approach described in Sect. 5.2.1.

5.3.1 Fitting a Weighted Power (or Quasi-Arithmetic) Mean

Power means involve a transformation of the inputs along with an inverse operation at the end so that we want each observed y_i to be modelled by

$$y_i = \left(\sum_{j=1}^{n} w_j x_{i,j}^p \right)^{\frac{1}{p}} \tag{5.1}$$

Modelling our objective function using the sum of absolute deviations would hence involve a transformation of the expression

$$\text{Minimize} \sum_{i=1}^{m} \left| \left(\sum_{j=1}^{n} w_j x_{i,j}^p \right)^{\frac{1}{p}} - y_i \right|,$$

which can not be converted into a linear program as we were able to with the
weighted arithmetic mean. However we can solve the problem approximately by
transforming all our evaluations using $f(t) = t^p$, since raising both sides of Eq. (5.1)
to the power p results in the right hand side inverse function cancelling out and
hence:

$$y_i^p = \sum_{j=1}^{n} w_j x_{i,j}^p.$$

The resulting objective for our linear program can hence be expressed as:

$$\text{Minimize} \sum_{i=1}^{m} \left| \left(\sum_{j=1}^{n} w_j x_{i,j}^p \right) - y_i^p \right|.$$

In practice, this means we can simply transform all the data (inputs and observed
outputs) using the appropriate power p, and then use linear optimization in the same
way we did for the weighted arithmetic mean. In fact we can use this approach for
any quasi-arithmetic mean, transforming the data using our generator $g(t)$.

It should be re-emphasized that this is not an *exact* solution to the problem, since
in terms of our residual values we actually would have

$$(PM_{\mathbf{w}}(\mathbf{x}_i) - r_i^+ + r_i^-)^p = y_i^p.$$

Depending on the data distribution we may end up overestimating or underestimat-
ing the true residuals in the fitting process (see [1] for approaches to improving the
accuracy).

5.3.2 Fitting an **OWA** Function

OWA functions are calculated similarly to the WAM and can be fit in a similar way.
The difference of course being that OWA weights are assigned based on relative
size of the inputs. This condition is fairly easy to account for in the linear fitting
process—we simply need to rearrange the data so that w_1 is fit to the lowest input,
w_2 is fit to the second lowest input and so on. Once the dataset is rearranged so that
the observed inputs for each instance are in non-decreasing order, the weights can
be found using the same linear programming approach as explained in Sect. 5.2.1.

5.3.3 Fitting the Choquet Integral

As might be expected, the approach to identifying the fuzzy measure weights for a Choquet integral can be somewhat more complicated. One approach is to consider an alternative representation of the Choquet integral.

$$C_{\mathscr{M}}(\mathbf{x}) = \sum_{A \subseteq \{1:n\}} \mathscr{M}(A) \min_{j \in A} x_j$$

where $\mathscr{M}$ is the Möbius transform of the fuzzy measure given by

$$\mathscr{M}(A) = \sum_{B \subseteq A} (-1)^{|A \backslash B|} v(B).$$

Notation Note Minimum of a subset of values

The $\min_{j \in A} x_j$ term in the function means that indicates that we look at all the inputs in the set A and take the smallest one. For example, if $\mathbf{x} = \langle 0.3, 0.5, 0.9, 0.4, 0.1 \rangle$ and $A = \{1, 3, 4\}$ then we would take the minimum of $x_1 = 0.3, x_3 = 0.9$ or $x_4 = 0.4$. This would hence be 0.3, which we would multiply by the Möbius measure of the set A.

The Möbius transform includes a calculation of $(-1)^{|A \backslash B|}$ which means -1 to the power of the cardinality of set A after set B is removed.

The Möbius transform is invertible, so we can actually learn the weights using the Möbius representation of a fuzzy measure and then convert them into our standard fuzzy measure weights v,

$$v(A) = \sum_{B \subseteq A} \mathscr{M}(B).$$

In order to prepare the dataset for data fitting, for each instance we calculate the $\min_{j \in A} x_j$ corresponding with each of the 2^n subsets. The fitting can now proceed similarly to how it would with finding w_1, w_2, etc for a weighted arithmetic mean, however since Möbius weights can be negative, we need to specify alternative linear constraints to ensure that $v(\{1 : n\}) = 1$ and that the output values for v are monotone with respect to adding additional elements to a set. These are given as:

$$\sum_{B \subseteq A | j \in B} \mathscr{M}(B) \geq 0, \text{ for all } A \subseteq \{1 : n\} \text{ and all } j \in A,$$

$$\mathscr{M}(\emptyset) = 0, \qquad \sum_{A \subseteq \{1:n\}} \mathscr{M}(A) = 1.$$

> **Notation Note** All subsets B that include j
> The subscript in the monotonicity constraint $B \subseteq A | j \in B$ means that for a given A, we add a constraint for each element in A where we add up all the subsets that include j. For example, for $A = \{2, 3, 5\}$, the constraint we would add for 3 would be the sum $\mathscr{M}(\{2, 3\}) + \mathscr{M}(\{3\}) + \mathscr{M}(\{3, 5\}) + \mathscr{M}(\{2, 3, 5\}) \geq 0$, and similarly for $j = 2, 5$.

Choquet integrals will have $2^n - 2$ unknown parameters for the fuzzy measure and include a number of monotonicity constraints however they can be learnt from data using linear programming in an analogous way to the WAM with the r_i^+, r_i^- residuals. Since increasing the number of variables will exponentially increase the size of the constraints array, some simplification assumptions on the fuzzy measure have been introduced so that the Choquet integral can still be used (e.g. k-additive measures [6]).

5.4   Using Aggregation Models for Analysis and Prediction

After learning the parameters for our chosen aggregation model, we can use the model for any of the following:

- Compare different averaging functions in terms of how well they model the data;
- Make inferences about the 'importance' of each variable based on the fitted weighting vector (or fuzzy measure);
- Make inferences about whether the relationship between the input variables and the output is one tending toward lower or higher inputs.
- Predict the outputs for unknown/new data;

We will work through an example applying the methods we have used for analysing data relating to bicycle share systems (available from the UCI Machine Learning Repository [7]). A number of cities across the USA (and other countries) have implemented bicycle share systems where users can rent a bike from a number of small parking stations. The systems are particularly appealing for tourists who want to use a bicycle to get around the city, as well as for those living in the city who want to make casual use of a bike without having to own and maintain one. The number of casual and registered users renting a bike from these systems each hour has been collected and stored at [4]. The authors of [5] matched this data against weather and seasonal information. We use a subset of the data consisting of 231 entries recording information relating to bikes rented over the weekend between 11am–12pm on different days and locations.[1] The variables included are as follows:

Weather   Categorical variable

- 1: Clear, Few clouds, Partly cloudy
- 2: Mist + Cloudy, Mist + Broken clouds, Mist + Few clouds, Mist

---

[1] Available at http://www.researchgate.net/publication/306099814_AggWAfit_R_library or alternatively, http://aggregationfunctions.wordpress.com/book.

- 3: Light Snow, Light Rain + Thunderstorm, Light Rain + Scattered clouds
- 4: Heavy Rain + Ice Pellets + Thunderstorm + Mist, Snow + Fog

Temperature   Temperature in Celsius
Humidity   Humidity (given as a percentage)
Windspeed   Windspeed (measured in km/h)
Casual users   Count of casual users that used a bike at that time.

Our aim is to understand more about the relationship between the weather conditions and the number of casual users that choose to rent a bike. We would expect that nicer weather conditions should be more favorable for bike riding and lead to more people deciding to rent bikes. The first ten entries are provided in the table below to give an idea of its structure.

Weather	Temperature (C)	Humidity	Windspeed	Users
1	6	81	19	26
2	6	71	17	16
2	−4	69	26	2
1	−8	40	35	2
1	−3	55	15	18
1	−1	44	17	23
2	−11	38	9	4
1	−7	43	15	22
2	−3	64	9	4
1	2	75	0	9

Before we perform fitting we need to first transform the data so that it is appropriate to use with aggregation functions. We can use scatterplots to obtain a snapshot of the approximate relationship between each of the weather information variables and the number of casual users (Fig. 5.3).

Although *weather* is categorical, we can see that as the class of weather becomes worse, the number of casual users tends to go down. Although there are four potential labels, there was no data instance that was recorded as having 'Heavy rain + Ice Pallets + Thunderstorms ...'. A simple negation and transformation to the unit interval (indicating suitability for bike riding) can be achieved using

$$f(t) = \frac{3 - t}{2},$$

so that good weather is allocated a score of 1, mist and cloudy is allocated 1/2, and snow and rain has a score of 0.

The *temperature* shows a positive linear trend, although for the few observations of temperatures above 35, the number of users does seem to decrease. We can use a utility transformation where we assume 30° is our 'ideal' temperature, and then assign the utility/suitability score based on how far we are from this, i.e. given that the minimum temperature recorded is −11,

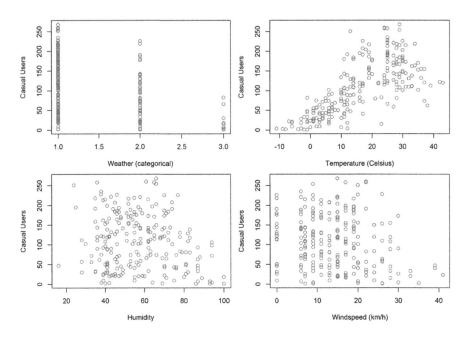

Fig. 5.3 Scatterplots for each of the predictor variables against the number of casual users between 11am and 12pm

$$f(t) = 1 - \frac{|t - 30|}{41}.$$

A temperature of 30° is allocated a score of 1, while a temperature of either 35 or 25 will be $1 - \frac{5}{41} \approx 0.88$.

There does not seem to be a strong relationship between *humidity* and the number of casual users, however it could be the case that humidity only has an impact when the temperature is high, so it still may provide some useful information to our models. We will again use a transformation according to suitability and assume (based on the high numbers in the plot) that 60 % represents 'an ideal' level.[2] Incorporating the minimum recorded humidity of 16, we will transform the values using

$$f(t) = 1 - \frac{|t - 60|}{44}.$$

A humidity of 60 will have a suitability score of 1, while a humidity of 20 would be transformed to 0.09.

---

[2]Of course, this may not be the case at all—we are only going by a very rough estimate of patterns in the data.

The *windspeed* shows an approximate negative trend. This is what we might expect, given that riding in high winds might be fairly uncomfortable. The highest windspeed is 41, so to apply a basic negation and scaling to the unit interval we can use

$$f(t) = 1 - \frac{t}{41}.$$

Finally, the number of casual users can be transformed using linear feature scaling with the maximum number of users recorded at 268 and the minimum equal to 1,

$$f(t) = \frac{t - 1}{267}.$$

For R implementation of the above transformations, we can use the following code to load the data transform the variables.

```
bike.data <- read.table("BikeShare231.txt")
bike.data[,1] <- (3-bike.data[,1])/2
bike.data[,2] <- 1-abs(bike.data[,2]-30)/41
bike.data[,3] <- 1-abs(bike.data[,3]-60)/44
bike.data[,4] <- 1-bike.data[,4]/41
bike.data[,5] <- (bike.data[,5]-1)/267
```

Plots of the transformed data are shown in Fig. 5.4.

5.4.1 Comparing Different Averaging Functions

After applying appropriate transformations, we can use the linear programming techniques presented in this chapter to learn the optimal weights for a given aggregation function. In some contexts, we may have a specific aggregation function that we want to use for our model, however in other cases it can be useful to compare different models and see which is the best.

Included in the AggWAfit R library[3] are functions that implement the least absolute deviation approach to finding weights.

fit.QAM() This function can be used to find weights for any power mean (including the weighted arithmetic, geometric mean, etc). The input is a data matrix with m rows and $n + 1$ columns (n is the number of variables and the observed y values should be in the last column) and produces two output files. One is the dataset along with the predicted y values appended as the final column,

---

[3]Available from http://www.researchgate.net/publication/306099814_AggWAfit_R_library or alternatively http://aggregationfunctions.wordpress.com/book.

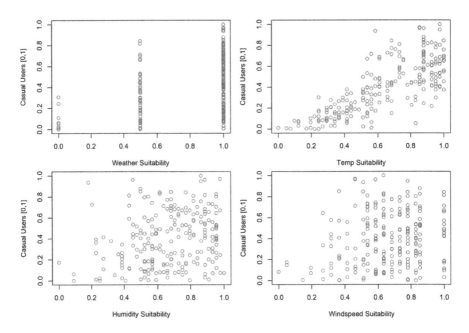

Fig. 5.4 Scatterplots of the BikeShare data after applying transformations

while the other is a 'stats' file, which gives the fitted weighting vector along with the RMSE, average absolute error, Pearson correlation and Spearman correlation. Optional inputs include alternative filenames for the output and stats files (the defaults are "output1.txt" and "stats1.txt"), and generator functions that change the aggregation function whose weights are to be identified. For example, to fit a geometric mean to our data, which we assigned to a matrix A), with output files "my-gmean.txt" and "gmean-stats.txt", we would use:

`fit.QAM(A,"my-gmean.txt","gmean-stats.txt",GM,invGM)`

Some generators are included in the `AggWAfit` library, however any additional generators can be added. They need to be defined separately and in terms of a single input. For example, the generators for a power mean with $p = 4.6$ could be defined using:

`PM4.6 <- function(x) {x^4.6}`
`invPM4.6 <- function(x) {x^(1/4.6)}`

`fit.OWA()` This fits the weights for an OWA operator with the same input options as `fit.QAM`, however without the generator options. In the output it will also provide the orness measure.

`fit.choquet` This function fits the fuzzy measure associated with a Choquet integral. In the stats file it provides the values of the fuzzy measure in binary ordering and also provides the Shapley indices. It includes an additional parameter relating to the k-additivity (which allows simpler models graduating between the arithmetic mean when $k = 1$ and the standard Choquet integral when $k = n$). We will leave this option set to its default.

These functions are all based on minimizing the least absolute deviation between the supplied y_i values and those predicted by the function. The R code first sets up the objective function and constraints, and then uses the lp() function from the lpSolve package to find the optimal decision variables (our weights). The time it takes to find the optimum weights is usually quite fast, however the total time will depend on the number of instances and the number of variables (i.e. the number of rows and columns of the matrix supplied).

Remember that in the case of fuzzy measures, there are $2^n - 2$ weights to be identified, so once we have 10 input columns we already are dealing with the equivalent of solving for a weighted arithmetic mean with over 1000 columns. This is not just a problem in terms of the time it takes, it also has implications for the number of data that would be required to avoid overfitting.

We will first use the fit.QAM() function to learn weights for a number of power means. The default parameters will fit a weighted arithmetic mean.

```
fit.QAM(bike.data)
```

This produces our output and stats file. Opening up the stats file gives the following information.

```
RMSE 0.295330008744615
Av. abs error 0.251528320040241
Pearson correlation 0.769578196865008
Spearman correlation 0.785917357818697
i w_i
1 0
2 1
3 0
4 0
```

This tells us that the best choice of weighting vector is to allocate all the weight to w_2 (temperature). The RMSE and average absolute error show a relatively high error, i.e. on average the prediction is out by about 25 %. Although we might expect that temperature would be the best predictor, given that all of our variables show a roughly positive relationship, we would expect that a combination should lead to better performance. Repeating the fitting process using the fit.OWA and fit.choquet functions helps us to identify the problem.

Fitting the OWA:

```
RMSE 0.205083254631973
Av. abs error 0.164581964757604
Pearson correlation 0.641633710128862
Spearman correlation 0.698185686615564
Orness 0
i w_i
1 1
2 0
3 0
4 0
```

Fitting the Choquet integral:

```
RMSE 0.19849960645762
Av. abs error 0.160854053748119
Pearson Correlation 0.695576455873277
Spearman Correlation 0.746909882563355
Orness 0.0518622555139716
i Shapley i
1 0.237411568872188
2 0.309821057012106
3 0.238972118185676
4 0.213795255930045
binary number fm.weights
1 0
2 0
3 0.141697877652858
4 0
5 0
6 0.144818976279836
7 0.144818976279836
8 0
9 0
10 0
11 0.141697877652858
12 0
13 0
14 0.144818976279836
15 1.00000000000001
```

A first look at our results suggests that something might have gone wrong.

(Thinking Out Loud)

The OWA allocates all the weight to the smallest input, while the Choquet integral has an orness of close to zero (we will look more closely at how to read the fuzzy weights later on—the last weight is slightly above 1 due to rounding). This indicates that both functions are tending heavily toward the minimum, which could be indicative of our models over-predicting the outputs. A potential cause of this is a skewed distribution of the outputs compared to the inputs. Recall that our models will all be idempotent, so if the suitability of each of the variables is 0.6 then the result should be 0.6 as well. Looking more closely at our transformed data in Fig. 5.4, we might have noticed that most of the data points would fall below a diagonal line going from $(0, 0)$ to $(1, 1)$ for each of the variables.

We can apply a further transformation to our casual users variable to alleviate this. We will use $f(t) = \sqrt{t}$.

```
bike.data[,5] <- sqrt(bike.data[,5])
```

We can see the difference this makes for the temperature variable in Fig. 5.5, which would be similar for all four predictor variables.

In the hope that this has fixed our problem, we can now repeat the fitting process. This time we will use fit.QAM for a number of generators so that

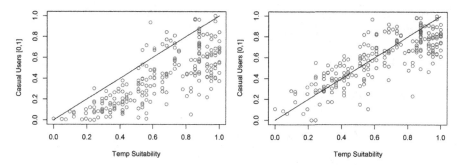

Fig. 5.5 Scatterplot showing relationship between casual users and temperature with the transformations used originally (*left*) and after applying an additional $f(t) = \sqrt{t}$ transformation to the casual users variable (*right*)

we can find the weights for power means with $p = 0$ (the geometric mean,[4] fit by including the additional arguments g=GMa,g.inv=invGMa), $p = 0.5$ (g=PM05,g.inv=invPM05), $p = 1$ (weighted arithmetic mean, we can use the default settings) and $p = 2$ (g=QM,g.inv=invQM), we summarize the results in the following table (to 4 decimal places).

p	(GM) $p = 0$	$p = 0.5$	(WAM) $p = 1$	$p = 2$
RMSE	0.1442	0.1423	0.1498	0.1596
Av. abs error	0.1125	0.1138	0.1189	0.1267
Pearson	0.8255	0.8324	0.8134	0.7910
Spearman	0.8050	0.8122	0.7972	0.7831
w_1	0	0.0957	0.0406	0
w_2	0.7688	0.6974	0.7543	0.8008
w_3	0.2085	0.1146	0.1169	0.1318
w_4	0.0227	0.0923	0.0882	0.0674

Fitting for the OWA now gives:

```
RMSE 0.158787002830088
Av. abs error 0.126546078672212
Pearson correlation 0.72365780710627
```

---

[4] The geometric mean has generators $g(t) = -\log t$ and $g^{-1}(t) = e^{-t}$ (GM and invGM in AggWAfit), however since there are a number of 0s in our data, the transformed data will include infinite values and the linear program won't solve correctly. We instead approximate the geometric mean using generators $g(t) = t^{0.00001}$ and $g^{-1}(t) = t^{\frac{1}{0.00001}}$. Alternatively, for functions like the geometric mean or the harmonic mean, we could fit to the dataset with a small amount added to every value, e.g. fit.QAM(bike.data+0.0000001,g=HM,g.inv=invHM). Recall, however that this is not technically allowed since these functions aren't translation invariant.

```
Spearman correlation 0.709523640986703
Orness 0.284894032126309
i w_i
1 0.543093717212378
2 0.172539896280975
3 0.170956959421992
4 0.113409427084656
```

and fitting for the Choquet integral results in:

```
RMSE 0.126098652373691
Av. abs error 0.0963523886577514
Pearson Correlation 0.83749129360049
Spearman Correlation 0.816233158458288
Orness 0.292598093621075
i Shapley i
1 0.210833619801679
2 0.488793332912998
3 0.158930261877987
4 0.141442785407334
binary number fm.weights
1 0.135288650805669
2 0.0207101073300679
3 0.585906040991734
4 0
5 0.135288650805669
6 0.555038111104102
7 0.65380243534751
8 0
9 0.135288650805669
10 0.585906040991735
11 0.585906040991735
12 0
13 0.135288650805669
14 0.648562908372979
15 0.999999999999997
```

Overall we can see that the Choquet integral has the best fitting performance, however this is not surprising since it is defined with respect to 16 parameters rather than 4 and so is more flexible. All of the models can be interpreted as having an average error of between 11–16 %. If we knew the expected weather, temperature, humidity and windspeed for a particular day, we could hope that our model would predict the number of casual users within about 15 % (provided our model holds for unseen data). For predictions corresponding with 20 casual users, this amounts to about 3 users. Out of the power means we built models for, a value of $p = 0.5$ had slightly better performance in terms of RMSE and correlation, however the geometric mean was better in terms of absolute error. The OWA did not fit as well as some of the means, from which we can infer that the source of the inputs is more important for making predictions than their relative size.

> **R Exercise 23** *Load the "BikeShare231.txt" data and after performing each of the transformations, verify that you can obtain the same fitting results.*

We will now look more closely at the fitted weights.

5.4.2 Making Inferences About the Importance of Each Variable

From the weights for the weighted power means, we see that *temperature* was by far the most important variable for making predictions, followed by the humidity, windspeed and then weather category. Models that fit worse had slightly higher allocations of weight to temperature, since as weights become closer to 1 the function behavior will be less influenced by the value of the parameter p.

Looking at the Shapley indices for the Choquet integral, however we see that relatively more importance was allocated to weather suitability than in the power mean models. Recall that the values of the fuzzy measure are given in binary order. So the 6th entry, which is 0110 in binary, will correspond with $v(\{2, 3\})$. The weights are as follows.

$$v(\{1, 2, 3, 4\}) = 1$$

$$v(\{1, 2, 3\}) = 0.654 \quad v(\{1, 2, 4\}) = 0.586 \quad v(\{1, 3, 4\}) = 0.135 \quad v(\{2, 3, 4\}) = 0.649$$

$$v(\{1, 2\}) = 0.586 \quad v(\{1, 3\}) = 0.135 \quad v(\{1, 4\}) = 0.135$$

$$v(\{2, 3\}) = 0.555 \quad v(\{2, 4\}) = 0.586 \quad v(\{3, 4\}) = 0$$

$$v(\{1\}) = 0.135 \quad v(\{2\}) = 0.021 \quad v(\{3\}) = 0 \quad v(\{4\}) = 0$$

$$v(\emptyset) = 0$$

The main thing we can observe with the fuzzy measure weights is that although temperature by itself ($v(\{2\})$) is low, once we combine it with any of the other inputs we have a subset weight of above 0.5. Humidity and windspeed are both worth nothing either by themselves or as a pair, and combining either or both with weather does not increase the weight of the measure. An overall interpretation of this fuzzy measure is that we need both suitable temperature, *and* additional factors in order for there to be a high number of users.

5.4.3 Make Inferences About Tendency Toward Lower or Higher Inputs

With regard to the OWA weights, be careful to note that these correspond with the relative size of each input, not the source of the input. The first weight, $w_1 = 0.5431$, indicates that more than 50 % of the weight is allocated to the smallest input. Similar to our interpretation of the Choquet integral, this suggests that many of the inputs should be in the suitable range before we are likely to see a high number of casual bike users.

The orness values of close to 0.3 (for both the Choquet model and the OWA) summarizes this tendency toward the lower inputs. If the orness were high, then it would mean that only a few of the factors would need to be high to obtain a high output overall, e.g. a comfortable humidity might correspond with a high number of users, even if the temperature were low and there were high winds. In multi-criteria decision problems, functions with low orness will favor alternatives that satisfy more of the criteria.

We can also make some inferences about the data tending toward high or low inputs by looking at the goodness-of-fit measures for the power means. In this case, lower values of p (i.e. functions tending toward lower inputs) tended to have better fitting accuracy than when p was higher.

5.4.4 Predicting the Outputs for Unknown/New Data

Once we have selected the model that seems most suitable, we can then use the parameters we identified in order to predict the output for an unknown instance. Suppose the weather next Saturday is forecast to be partly cloudy, with a temperature of 18 °C, humidity at 63 and a windspeed of 12 km/h. What would each of the models predict for the number of users?

First, we need to transform these values as we had previously. The raw input vector is $\langle 1, 18, 63, 10 \rangle$, which after applying each of our transformations:

$$x_1' = \frac{3-1}{2}, \quad x_2' = 1 - \frac{|18-30|}{41}, \quad x_3' = 1 - \frac{|63-60|}{44}, \quad x_4' = 1 - \frac{12}{41}$$

results in the vector,

$$\mathbf{x} = \langle 1, 0.7073, 0.932, 0.756 \rangle.$$

Entering this vector and the weights for each model into R produces the outputs in the table below, which we again need to transform in order to properly interpret. For the number of casual users, we subtracted 1, divided by 267 and then took the square root. To undo this, we therefore need to take the square, multiply by 267 and then add 1.

Function	GM	PM $p = 0.5$	WAM	PM $p = 2$	OWA	Ch
Model output	0.7503	0.7623	0.7497	0.7441	0.7873	0.7469
Casual users	151.31	156.16	151.09	148.82	166.50	149.95

Depending on the model we use, we would predict there to be about 150–165 users on the coming Saturday. The OWA gives a higher result than the other functions because it is the only one that doesn't take into account the source of the inputs. In this case, the temperature has the lowest transformed value, and so the other functions all end up giving a lower prediction.

5.5 Reliability

If we want to be able to make inferences about the importance of variables or use our functions for prediction, we need to be able to assess the reliability of our models. Here we differ from the usual statistics approach, which is to look at hypothesis tests and confidence intervals, and focus more on whether our model is reasonable. Some questions we might initially consider are:

- Do our y_i values represent a 'ground truth', or are they a proxy measure themselves (or potentially erroneous/noisy in some other way)?
- Are we evaluating an approach or the model itself?

We'll look at each of these questions in turn.

5.5.1 Do Our y_i Values Represent the 'Ground Truth'?

If we intend to make claims about the relationship between the inputs and outputs, or even about the general usefulness of a certain aggregation function for modelling a particular dataset, we need to be clear about what it means for the model to be accurate. If the y_i observations represent human judgements, we might expect these to have approximation errors or inconsistencies, and a perfectly fitting function may not be what is desired.

Similarly, it may be great that we have an aggregation model that can make predictions based on our training dataset, however if our training dataset is based on 'expert' evaluations that incorporate prior assumptions about relationships between inputs and outputs, then the predicted values for unseen observations may also reflect these biases.

There are two basic types of datasets we can use for performing experiments: synthetic and real. Both have their own drawbacks.

- Synthetic

  - Need to be generated in a theoretically sound way appropriate to the problem;
  - Always carry some skepticism—e.g. what biases and distributions might exist when it comes to real data that would not be captured by the dataset we generated?

- Real

  - Difficult to obtain due to permission, privacy concerns, availability, or even existence;
  - Can be incomplete;
  - Can involve variables that are difficult to transform to numerical data;
  - Can be too small or case specific.

Creating plausible synthetic data can be difficult. We need to think carefully about what kind of distribution the data should have (and whether particular distributions might cause problems). We need methods for generating natural noise or errors that we would be likely to have with a real dataset. The type of noise we have, or the generation of the model, can often influence which functions will be robust in the first place. For example, if the variables are generated independently of one another and the 'errors' added to the ideal model are normally distributed, then it is likely that the weighted arithmetic mean will perform the best.

For real data, we need to ensure the data is complete (or have a way of suitably filling in the missing values) and does not contain errors. Before performing any transformations, we might want to make sure we understand the units and the context in which the data was collected (e.g. measurements, observations, surveys etc).

There are countless small errors in reasoning that can mean either real or synthetic datasets become useless.

5.5.2   Are We Evaluating an Approach or the Model Itself?

We can look to build aggregation models from data for multiple reasons. On the one hand, we may have data where we know (or have an intuition) that one variable in some way summarizes the others or is dependent on the others, and we want to understand the nature of this relationship. For example, can movie box office takings be predicted from an aggregation of social media analytics? Can tumors be classified as benign or malignant based on texture features from MRI scans? Can we predict whether a potential employee will perform well in our company? In such cases, we might be more interested in the parameters of the model itself, i.e. which analytics are the most important and useful? Which texture features or other indicators are most important when it comes to classifying tumors? What are the characteristics when we are looking for new employees? We will want the functions to fit very closely in order to make reliable interpretations and we will want to ensure that we have enough data to safely say that the model is stable.

On the other hand, we may be looking at the use of aggregation functions as a general approach to making predictions in a given domain. We therefore might be interested in a number of factors that could influence the overall fitting performance. For example:

- How does the reliability change with the number of training data? i.e. are there significant differences in either the fitting accuracy or prediction accuracy (for unseen data) when we use 10, 20, 50, 100 or 200 training data?
- How does the reliability change with the number of variables used? Is it best to use all the variables in the models, or are certain combinations better in terms of producing reliable results?
- How important are the transformations we made in determining the fitting or prediction accuracy?
- How does the use of aggregation functions compare with other approaches (e.g. from broader machine learning, we could compare with statistical regression, neural networks, decision trees)? The performance of aggregation functions might be worse but the advantage is their interpretability.

5.6 Conclusions

In the first chapter, some important traditional means were introduced: the harmonic mean, the geometric mean and the arithmetic mean, as well the median. We saw that in some cases, use of the arithmetic mean is not theoretically justified for giving the average of the data. This is also where we introduced some definitions and properties of aggregation functions. In the second chapter, we focused on potential ways of transformation data so that the variables exhibit a monotone relationship with the output and are given over the same scale. The power means were then introduced—a family of aggregation functions that includes the arithmetic, harmonic and geometric means as special cases. In Chap. 3 we showed how weights could be incorporated into the aggregation process in order account for the differing importance of each variable. We noted that sometimes the behavior with respect to weights and tendency toward high or low values could make it difficult to determine local behavior, i.e. which input should be changed to achieve the biggest increase. In Chap. 4 we introduced the ordered weighted averaging (OWA) operator and the discrete Choquet integral. We saw that these functions could also be characterized as tending toward higher and lower values and allocated weights depending on the relative order of variables. Although its definition is somewhat complicated, we saw that the Choquet integral could be used to model positive and interaction effects between variables. In this final chapter, we have introduced some methods for weights identification, and applied the concepts and understanding from previous chapters to the interpretation of models and data.

Of course, when working with real world data there are many more problems
that can arise (and many more potential methods to address them) that are beyond
the scope of this text. It is worth reiterating that a key advantage of aggregation
functions over some other techniques lies in the sound theoretical understanding
developed across various disciplines that allows their clear and meaningful interpretation. Even as new and powerful techniques are being introduced, an understanding
of how to analyze data with aggregation functions, as well as being useful in of itself,
can provide a sound basis for interpreting more complex methods and models.

5.7 Summary of Formulas

Minkowski Distance

$$d_p(\mathbf{x}, \mathbf{y}) = \left(\sum_{i=1}^{n} |x_i - y_i|^p \right)^{1/p} \tag{5.2}$$

Cosine Similarity

$$\mathrm{Sim}_{\cos}(\mathbf{x}, \mathbf{y}) = \frac{\sum\limits_{i=1}^{n} x_i \cdot y_i}{\left(\left(\sum\limits_{i=1}^{n} x_i^2 \right) \cdot \left(\sum\limits_{i=1}^{n} y_i^2 \right) \right)^{1/2}} \tag{5.3}$$

Fitting WAM Weights Linearly

$$\text{Minimize} \quad \sum_{i=1}^{m} r_i^+ + r_i^- \tag{5.4}$$

$$\text{s.t.} \left(\sum_{j=1}^{n} w_j x_{i,j} \right) - r_i^+ + r_i^- = y_i, \qquad i = 1, 2, \ldots, m$$

$$\sum_{j=1}^{n} w_j \quad = 1$$

$$w_j \quad \geq 0, \qquad j = 1, 2, \ldots, n$$

$$r_i^+, r_i^- \quad \geq 0$$

RMSE

$$\mathrm{RMSE} = \sqrt{ \sum_{i=1}^{m} \frac{(A_{\mathbf{w}}(\mathbf{x}_i) - y_i)^2}{n} } \tag{5.5}$$

Av. AE

$$\text{Av.AE} = \sum_{i=1}^{m} \frac{|A_w(\mathbf{x}_i) - y_i|}{n} \tag{5.6}$$

Pearson Correlation

$$r = \frac{\sum_{i=1}^{m} (A_w(\mathbf{x}_i) - \bar{A})(y_i - \bar{y})}{\sqrt{\sum_{i=1}^{m} (A_w(\mathbf{x}_i) - \bar{A})^2} \sqrt{\sum_{i=1}^{m} (y_i - \bar{y})^2}} \tag{5.7}$$

(Spearman correlation is the same except with x_i, y_i replaced with ranks)

5.8 Practice Questions Using R

1. The Kei Hotels rating data is a 56×10 table where the first column indicates Kei's ratings for each hotel (out of 100) and columns 2–9 are the ratings of similar users.

   (i) Download the `KeiHotels.txt` file and save it to your R working directory.
   (ii) Assign the data to a matrix, e.g. using
   ```
   kei.data <- as.matrix(read.table("KeiHotels.txt"))
   ```
   (iii) Define a function to measure the similarity between Kei and the other online users. (The Euclidean distance can be defined using the Minkowski distance with $p = 2$)
   (iv) Which of the users is *most similar* to Kei? Investigate using scatterplots, histograms, correlation and similarity between Kei and the other users.

2. Download the `AggWaFit.R` file to your working directory and load into the R workspace using,
   ```
   source("AggWAfit.R")
   ```

   (i) Using `fit.QAM`, find the weights for a weighted arithmetic mean that best approximates Kei's ratings from those of the other users.
       [hint: You will need to set Kei's data as the last column. You can do this using (e.g. if your data matrix is 'A'), `A <- A[,c(2:9,1)]`]
   (ii) Use `fit.QAM`, `fit.OWA` to find the best weights for:

   - Weighted power means with $p = 0.5$, and $p = 2$, (the generators required are PM05, invPM05 and QM, invQM).
   - A geometric mean (the generators are GMa and invGMa).
   - An OWA.

   You can also experiment with using only a subset of the variables.

(iii) Which model fits the data the best?

(iv) Comment on similarities and differences between the users that were found to be the most similar to Kei and whether they had the highest weights allocated in the fitted data models.

3. Use a subset of any four of the similar users and the `fit.choquet` function to find the fuzzy measure that fits the data best and compare with your previous findings (trying to use more users will result in a very long time to find the values).

References

1. Bartoszuk, M., Beliakov, G., Gagolewski, M., James, S.: Fitting aggregation functions to data: part I Linearization and regularization. In: Information Processing and Management of Uncertainty in Knowledge-Based Systems (IPMU 2016), Eindhoven, Netherlands, pp. 767–779 (2016)
2. Beliakov, G., Pradera, A., Calvo, T.: Aggregation Functions: A Guide for Practitioners. Springer, Heidelberg (2007)
3. Berkelaar, M., et al.: lpSolve R package. Available online. https://cran.r-project.org/web/packages/lpSolve.Cited4August2016 (2015)
4. Capital Bikeshare: System Data. http://www.capitalbikeshare.com/system-data (2016). Cited 10 Aug 2016
5. Fanaee-T, H., Gama, J.: Event labeling combining ensemble detectors and background knowledge. Prog. Artif. Intell. 2, 1–15 (2013)
6. Grabisch, M.: k-order additive discrete fuzzy measures and their representation. Fuzzy Sets Syst. 92, 167–189 (1997)
7. Lichman, M.: UCI Machine Learning Repository. University of California, School of Information and Computer Science, Irvine, CA. http://archive.ics.uci.edu/ml (2013)
8. Linden, G., Smith, B. and York, J.: Amazon.com recommendations: Item-to-item collaborative filtering. IEEE Internet Comput. 7(1), 76–80 (2003)
9. Pearson, K.: Notes on regression and inheritance in the case of two parents. Proc. R. Soc. Lond. 58, 240–242 (1985)
10. R Core Team: R: A language and environment for statistical computing. R Foundation for Statistical Computing, Vienna. http://www.R-project.org/ (2014)
11. Spearman, C.: The proof and measurement of association between two things. Am. J. Psychol. 15, 72–101 (1904)
12. Turlach, B.A., Weingessel, A.: quadprog R package. Available online. https://cran.r-project.org/web/packages/quadprog.Cited4August2016 (2013)
13. Winston, W.L.: Operations Research: Applications and Algorithms, 4th edn. Cengage Learning, Belmont, CA (2003)

Chapter 6
Solutions

Chapter 1 Aggregating Data with Averaging Functions: Solutions

1.9   Practice Questions

1. Write down the arithmetic mean, the geometric mean and the harmonic mean explicitly for 3 and 4 arguments (i.e. in terms of x_1, x_2, x_3, x_4).

> For the arithmetic mean, explicitly we have
>
> $$\frac{x_1 + x_2 + x_3}{3} \quad \text{and} \quad \frac{x_1 + x_2 + x_3 + x_4}{4}$$
>
> For the geometric mean, we can write,
>
> $$(x_1 x_2 x_3)^{1/3} \quad \text{and} \quad (x_1 x_2 x_3 x_4)^{1/4}$$
>
> It would also be fine to write the arithmetic mean as $\frac{1}{3}x_1 + \frac{1}{3}x_2 + \frac{1}{3}x_3$ and the geometric mean as $x_1^{1/3} x_2^{1/3} x_3^{1/3}$, or use equivalent decimals etc. The main meaning of 'explicit' here is to remove the $\sum$ and index notation.

2. Using the concept of monotonicity, give some reasoning as to why it always holds that $G(\mathbf{x}) \geq \min(\mathbf{x})$.

> This will be true as long as none of the inputs are negative (and in many cases the geometric won't be defined if for negative numbers).
> If we have all the inputs x_i equal, e.g. $x_1 = x_2 = \cdots = x_n = a$ then we would have

$$\underbrace{(a \times a \times \cdots \times a)}_{n \text{ times}}^{1/n} = (a^n)^{1/n} = a.$$

If any of the inputs increase, the product inside the brackets must increase, because we are multiplying by a larger number. If we increase any number of the inputs, except for 1 (so one of the $x_i = a$ and this becomes the smaller number), then we will have had a series of increases and therefore the output must be greater than a (or equal to a if $a = 0$).

3. Are the harmonic mean and geometric mean homogeneous and translation invariant? How about the median?

The harmonic mean and geometric mean are homogeneous but not translation invariant. Multiplying the inputs by a constant factor will increase the output at the same rate, however adding values by a given value does not increase the output by the same value. On the other hand, since the median is the middle value and multiplication and addition of all inputs won't change that, the output will change in the same way the middle value is transformed. So the median is both translation invariant and homogeneous.

4. If $AM(23, 42, 27, 68) = 40$, without calculating, what will be the value of

$$AM(29, 48, 33, 74)?$$

All of the inputs have been increased by 6, so the output will change accordingly to 46 (due to translation invariance).

5. If $AM(x_1, x_2, x_3, x_4) = 15$, what will be the value of

$$AM(x_1 + 2, x_2 + 2, x_3 + 2, x_4 + 2)?$$

All of the inputs have had 2 added, so the output will be 17.

6. If $AM(x_1, x_2, x_3, x_4) = 341$, what will be the value of

$$AM(x_1 + c, x_2 + c, x_3 + c, x_4 + c)?$$

All of the inputs have had c added, so the output will be $341 + c$.

7. For the vector $\mathbf{x} = \langle 11, 21, 34, 30 \rangle$ the arithmetic mean is 24, the geometric mean is 22.03196, and the harmonic mean is 19.87348, what will be the values if $\mathbf{x} = \langle 110, 210, 340, 300 \rangle$?

Since all of these functions are homogeneous and the inputs have just been multiplied by 10, the outputs will also be 10 times greater. So the arithmetic mean will be 240, the geometric mean will be 220.3196, and the harmonic mean will be 198.7348.

8. If $GM(x_1, x_2, x_3, x_4) = 6$, what will be the value of

$$GM(3x_1, 3x_2, 3x_3, 3x_4)?$$

The inputs have been multiplied by 3, so the output will change accordingly to 18 (due to homogeneity).

9. If $AM(x_1, x_2, x_3, x_4) = 7$, what will be the value of

$$AM(3x_1 + 2, 3x_2 + 2, 3x_3 + 2, 3x_4 + 2)?$$

The inputs have been multiplied by 3 and 2 has been added. Since the arithmetic mean is translation invariant and homogeneous, this means the new output will be $3 \times 7 + 2 = 23$.

10. Which of the functions we've introduced so far are Lipschitz continuous?

The arithmetic mean, harmonic mean and median are all Lipschitz continuous. This means that changing one of the inputs slightly cannot have a big impact on the output of the function. The only function we've studied which is not Lipschitz continuos is the geometric mean.

11. Explain what it means to say that the arithmetic mean, harmonic mean and geometric mean could be affected by outliers.

The arithmetic mean can be affected by extremely high or extremely low inputs. For each change in the outlier, the output of the arithmetic mean will change at a rate of $\frac{1}{n}$. For example, if there are 100 inputs, then an extreme outlier increasing by 100,000 will increase the overall output by 1000.

The geometric and harmonic means are less affected by high outliers, but more affected by low outliers, since they are more sensitive to low inputs.

1.10 Practice Questions Using R

1. Suppose you have $\mathbf{x} = \langle 0.1, 0.2, 0.6, 0.9 \rangle$, Calculate

    (i) The arithmetic mean
    (ii) The geometric mean

(iii) The harmonic mean

(iv) The median

and compare the results.

> (i) 0.45 (ii) 0.322371 (iii) 0.225 (iv) 0.4
>
> The geometric mean and harmonic mean both tend toward the lower inputs 0.1 and 0.2, while the median and mean are both quite central. In terms of being 'representative', the harmonic mean and geometric mean are closer to the two closest inputs (although this mainly happens because they are low. If we had 0.85 and 0.9 instead of 0.6 and 0.9, none of the inputs would be very close to this mini "cluster".

2. Create 2-variate versions of the geometric mean and harmonic mean using `F <- function(x,y) {...}`, i.e. where the `x` and `y` are two numbers rather than vectors.

> This could be done in a number of ways. Example implementations would be:
> ```
> GM <- function(x,y) {sqrt(x*y)}
> HM <- function(x,y) {2/(1/x+1/y)}
> ```

3. Define the function $f(x, y) = \frac{x^2+y^2}{x+y}$, $(x + y \neq 0$ and $f(0, 0) = 0)$ and evaluate $f(0.3, 0.9)$, and $f(0.4, 0.9)$. Based on your results, can it be stated that f is *not* an aggregation function?

> To implement the function,
> ```
> not.agg <- function(x,y) {
> out <- 0
> if((x+y)>0) {out <- (x^2+y^2)/(x+y)}
> out}
> ```
> We then have:
> `not.agg(0.3,0.9) = 0.75`, and `not.agg(0.4,0.9) = 0.7461538`. The important thing to notice is that even though the 0.3 increased to 0.4, the function output decreased, and therefore it is not an aggregation function (fails the monotonicity property).

4. The rise in house prices in Australia's 8 major cities was 6.8 % over 2014 and 9.8 % over 2013. What is the average yearly increase? (which operator should be used?)

> We should use the geometric mean. With price increase we need to find the geometric mean of 1.068 and 1.098, which is 1.082896 and therefore the average yearly increase is about 8.3 %.

5. A car travels at 110 km/h on the way to a destination and 80 km/h on the way back. What is its average speed? (which operator should be used?)

We should use the harmonic mean (speed is distance over time so if the times stay the same we can use the normal average but in this case the distances stay the same and the times change). The average speed is 92.63 km /h (we travel at 80 km for longer so the average will be closer to this than to the 110).

6. Let $\mathbf{x} = \langle 25, 14, 39, 21, 51, 22 \rangle$. Compare the outputs of the arithmetic, harmonic, geometric means and the median. How do these values differ if the last input $x_6 = 22$ is replaced with an outlier $x_6 = 288$?.

The arithmetic mean gives 28.66667 without the outlier and 73 with the outlier.

The geometric mean gives 26.17525 without the outlier and 40.18393 with the outlier.

The harmonic mean gives 24.02392 without the outlier and 28.87827 with the outlier.

The median is 23.5 without the outlier and 32 with (it changes because it was one of the middle values that changed).

Since it is a high outlier, it has more of an impact on the average with the arithmetic mean than it does with the geometric and harmonic means. The harmonic mean is least affected (out of these 3) because it tends more towards the lower inputs.

7. Let $\mathbf{x} = \langle 189, 177, 189, 212, 175, 231 \rangle$. Compare the outputs of the arithmetic, harmonic, geometric means and the median. How do these values differ if the last input $x_6 = 231$ is replaced with an outlier $x_6 = 11$?.

The arithmetic mean gives 195.5 without the outlier and 158.8333 with the outlier.

The geometric mean gives 194.5273 without the outlier and 117.1145 with the outlier.

The harmonic mean gives 193.5984 without the outlier and 51.03252 with the outlier.

The median is 189 without the outlier and 183 with (it changes because the 231 was in the top half of numbers and then moved to the bottom half of numbers, hence shifting all the positions up by 1 so that what was considered the middle two inputs changed.

In this case it is a low outlier, and hence the geometric and harmonic means are much more drastically affected.

Chapter 2 Transforming Data: Solutions

2.11   Practice Questions

1. Consider the following wait and service times for 100 customers.

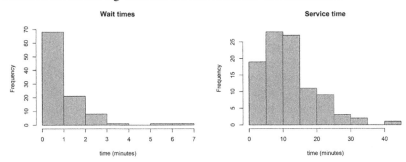

What would be some potential transformations that could be used to scale these variables to the unit interval?

The choice will depend on our purpose of aggregation. If we want to know the total time, then this is simply the wait and service time added together. However, perhaps we would like to evaluate the customer experience based on wait time and service time where each of these is considered equally important and contributes to an overall score. In this case, we need to apply scaling and negation (usually we'd say that a longer time is worse). We can also see from the distributions that both variables have more values closer to zero, so transformations like log, or x^p with p less than 1 could be useful. For example,

Wait Times Let the minimum be a and the maximum wait time be b. Denoting the original wait time by x and the transformed weight time by x':

$$x' = 1 - \frac{\ln x - \ln a}{\ln b - \ln a}$$

(this is equivalent to first transforming all the variables using log, then applying linear feature scaling, then applying the standard negation.)

Service Times Let the minimum be a and the maximum wait time be b. Denoting the original wait time by x and the transformed wait time by x':

$$x' = 1 - \frac{x^{1/2} - a^{1/2}}{b^{1/2} - a^{1/2}}$$

(Similar to wait times however using a square root instead of log since the data is not as skewed.)

2. Below are histograms of the accuracy and computation time taken for varying classification parameters. We want to be able to aggregate the two scores in order to be able to determine the best overall classifier.

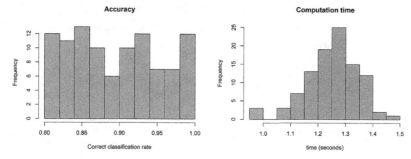

Suggest some transformations that would be able to transform the variables so that it makes sense to take their average.

Both of the variables are reasonably symmetric in terms of their distribution. For computation time, faster is better, so here we might use a negation and scale the data to the unit interval. Accuracy is already given over the scale 0 to 1. We could scale it so that a score of 80 % corresponds with something closer to zero if we wanted, or we could leave it as is. We can consider whether we would want a classifier with an accuracy of 0.8 and the fastest time to have an aggregated score of about 0.9 or of about 0.5. The following uses the latter.

Accuracy Let the minimum be a (and the maximum is 1). Denoting the original accuracy score by x and the transformed score by x':

$$x' = \frac{x - a}{1 - a}$$

Computation Time Let the minimum be a and the maximum be b, the original computation time is x and the transformed score for computation time will be x':

$$x' = 1 - \frac{x - a}{b - a}.$$

3. For the following piecewise function,

$$f(t) = \begin{cases} \frac{t}{5}, & t < 3, \\ \frac{3}{5} + 2\frac{t-3}{10}, & 3 \le t \le 5, \end{cases}$$

what will be the value when $t = 2$? How about when $t = 4$?

> When $t = 2$, we are in the domain $t < 3$, so that means we calculate $f(t)$
> using $\frac{t}{5}$. Therefore, when $t = 2$, the output value will be $2/5 = 0.4$.
> When $t = 4$, we are in the interval $3 \le t \le 5$. In this case, we use $\frac{3}{5} + 2\frac{t-3}{10}$.
> We will have
>
> $$f(4) = \frac{3}{5} + 2\frac{4-3}{10} = \frac{3}{5} + \frac{2}{10} = \frac{4}{5} = 0.8.$$

4. Is the power mean symmetric, homogeneous and translation invariant? Explain.

> **Symmetry** Since the power mean involves summing all of the inputs
> (after they are raised to the power p), the order of the inputs won't be
> important. All power means will therefore be symmetric.
>
> **Homogeneous** All power means are homogeneous (as stated in the topic
> text), however this can also be shown algebraically.
> (In case you are interested) For any λ, we will have:
>
> $$PM(\lambda x_1, \lambda x_2, \ldots, \lambda x_n) = \left(\frac{1}{n} \sum_{i=1}^{n} (\lambda x_i)^p \right)^{1/p}$$
>
> $$= \left(\frac{1}{n} (\lambda^p x_1^p + \lambda^p x_2^p + \ldots + \lambda^p x_n^p) \right)^{1/p}$$
>
> $$= \left(\frac{1}{n} \lambda^p (x_1^p + x_2^p + \ldots + x_n^p) \right)^{1/p} = \left(\frac{\lambda^p}{n} \sum_{i=1}^{n} x_i^p \right)^{1/p}$$
>
> $$= \lambda \left(\frac{1}{n} \sum_{i=1}^{n} x_i^p \right)^{1/p}$$
>
> So multiplying all of the inputs by λ is the same as aggregating first and
> then multiplying the output by λ.
>
> **Translation Invariance** It is only for $p = 1$ (i.e. the arithmetic mean)
> that the power mean will be translation invariant.

5. Does the power mean have absorbent elements? Explain.

> We already know that the geometric mean and harmonic mean (which are both special cases) have the absorbent element $a = 0$. This will be the case for any $p \leq 0$ (as stated in the topic text).
>
> (For those who are interested) The theory behind this is that if the inputs are raised to a negative power $-b$, we will have inputs transformed so that $x^{-b} = \frac{1}{x^b}$. This means that if $x = 0$, then the transformed input $\frac{1}{0}$ is included, which although undefined, is essentially ∞. When we take the inverse function, we will then have $\infty^{-1/b}$ or $\frac{1}{\infty^{1/b}}$, which is zero.

6. Write out the power mean explicitly for 3 arguments when $p = 2$.

> $$PM(x_1, x_2, x_3) = \left(\frac{x_1^2 + x_2^2 + x_3^2}{3} \right)^{1/2}$$
>
> or this can also be written,
>
> $$PM(x_1, x_2, x_3) = \sqrt{\frac{x_1^2 + x_2^2 + x_3^2}{3}}.$$

7. Write out the power mean explicitly for 2 arguments when $p = -4$.

> $$PM(x_1, x_2, x_3) = \left(\frac{x_1^{-4} + x_2^{-4}}{2} \right)^{1/-4}$$
>
> or this can also be written,
>
> $$PM(x_1, x_2, x_3) = \left(\frac{2}{\frac{1}{x_1^4} + \frac{1}{x_2^4}} \right)^{1/4}.$$

8. If $PM(9, 10, 17, 16) = 13$ for some value of p, can we work out the value of $PM(12, 13, 20, 19)$ and $PM(18, 20, 34, 32)$ without knowing p?

> We can only determine $PM(12, 13, 20, 19)$ if we know p. If $p = 1$ then translation invariance will mean that our output should be 16 (the arithmetic mean), however if $p \neq 1$ then this will not hold. Furthermore, in this case, we can tell that p is not 1 since the arithmetic mean of

> 9,10,17 and 16 should be 15.5. On the other hand, since the power mean is homogeneous for all p, we can deduce that the result should be 26 (double 13 since all the inputs are doubled).

9. If $p = -4$, can we determine the value of $PM(3, 0, 2, 8)$ without calculating?

> The output will be zero since $x_2 = 0$ and for negative values of p, the power mean has 0 as an absorbent element.

10. Write out the explicit formula (i.e. without the $\sum$ symbol) for $PM(x_1 + 3, x_2 + 3)$.

> We will have
> $$PM(x_1 + 3, x_2 + 3) = \left(\frac{(x_1+3)^p + (x_2+3)^p}{2} \right)^{1/p}.$$

11. If $PM(3, 7, 29, 45) = 7.162$, is the value of p likely to be greater than 1 or less than 1?

> This value of 7.162 seems closer to the two low inputs. Since the output would be 21 for $p = 1$, we can determine that p must be much lower.

12. If $PM(3, 7, 29, 45) = 36.258$, is the value of p likely to be greater than 1 or less than 1?

> This value is now closer to the higher inputs. Analogously to the previous question, we can tell that p must hence be higher than 1.

13. If $PM(2, 3, 8, 3) = 7$, is the value of p likely to be high or low?

> Since 7 is only relatively close to the 8, this power mean tends towards the highest input. That means that p must be high (it's close to 10).

2.13   Practice Questions Using R

1. Suppose you have $\mathbf{x} = \langle 0.3, 0.8, 0.1, 0 \rangle$, Calculate the power mean for the following cases

   (i) $p = 4$
   (ii) $p = 2.5$
   (iii) $p = 0$
   (iv) $p = -3.1$

   and comment on (i.e. compare) the results.

(i) 0.5684956,
(ii) 0.4758802,
(iii) 0 (special case of geometric mean—has absorbent elements)
(iv) 0.
  The higher value of $p = 4$ produces the higher output. Both $p = 0$ and $p = -3.1$ produce an output of zero since one of the inputs is zero.

2. Create a 2-variate function for the power mean when $p = 3$ using
   ```
   PM3 <- function(x,y) {...}.
   ```
   i.e. so that it takes the inputs x and y, which will be numbers rather than vectors.

   An example implementation would be
   ```
   PM3 <- function(x,y) {((x^3+y^3)/2)^(1/3)}
   ```

3. Load the data file "wait.service.txt" which has the wait and service times from Sect. 2.11 Question 1, and perform the following.

   (i) Use appropriate scaling techniques so that the two variables both have data given over the same range.

   Example steps would be
   Load the file
   ```
   S <- as.matrix(read.table("wait.service.txt"))
   ```
   To transform the first variable
   ```
   S[,1] <- 1- (S[,1] - min(S[,1]))/(max(S[,1]) - min(S[,1]))
   ```
   or alternatively, to account for the log distribution, first use
   ```
   S[,1] <- log(S[,1])
   ```
   and then apply the scaling transformation.
   For the second variable, similarly, we could first apply a transformation for the distribution, which could be
   ```
   S[,2] <- (S[,2])^(1/2)
   ```
   Then apply a negation and scaling,
   ```
   S[,2] <- 1- (S[,2] - min(S[,2]))/(max(S[,2]) - min(S[,2]))
   ```

   (ii) Calculate the output of the power mean for $p = -1, p = 0, p = 1$ and $p = 2$.

   To store these values, each time we can create a new empty vector.
   ```
   out.1 <- array(0,100)
   ```
   Then calculate all the values using our power mean function. For example, for $p = -1$,
   ```
   for(i in 1:100) out.1[i] <- PM(S[i,],-1)
   ```
   Assuming the transformations from (i) were used (the answers will differ slightly if no log and $x^{1/2}$ transformation were used, the first 5 values from the table would be:

For $p = -1$
| 0.4563546 | 0.5386007 | 0.4250787 | 0.4403588 | 0.0000000 |

For $p = 0$
| 0.5086135 | 0.5470779 | 0.4435698 | 0.4692176 | 0.0000000 |

For $p = 1$
| 0.5668566 | 0.5556886 | 0.4628652 | 0.4999677 | 0.2743486 |

For $p = 2$
| 0.6196493 | 0.5641679 | 0.4813877 | 0.5289331 | 0.3879875 |

(iii) What is the value of service and wait time that has the best aggregated value using each of the values of p?

To find these, we can find the entry leading to the highest input using the order function.
`order(out.1,decreasing = TRUE)`
The number given first will be the row of our data table that produced the best aggregated value.
In all cases, the best aggregated score was obtained for the 18th entry, which had original values
`0.008825986   3.558991131.`
The aggregated scores for each p were,
For $p = -1$ 0.9059214
For $p = 0$ 0.9099572
For $p = 1$ 0.9140111
For $p = 2$ 0.918047

(iv) Plot the outputs for each of the functions and compare the results.

We provide plots for $p = -1$ and $p = 2$ below. Note that the scales are different, however we can see that in general, the values for $p = 2$ are higher, with all outputs being above 0.2. We can identify the 18th entry as the highest in both cases, and for the value of $p = -1$ there are two values which have outputs of zero, however these do not appear to be among the lowest values for the case of $p = 2$.

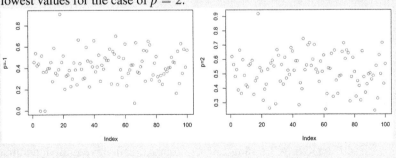

4. Load the two data files "comp.acc.txt" and "comp.time.txt" and, either merging them into a single table or keeping them as separate vectors, perform the following.

  (i) Use appropriate scaling techniques so that the two variables both have data given over the same range.

```
Example steps would be
Load the files
Comp1 <- as.matrix(read.table("comp.acc.txt"))
Comp2 <- as.matrix(read.table("comp.time.txt"))
Merge these together
Comp <- cbind(Comp1,Comp2)
To transform the first variable
Comp[,1] <- (Comp[,1] - min(Comp[,1]))/(max(Comp[,1])
- min(Comp[,1]))
Transform the second variable (need a negation here)
Comp[,2] <- 1 - (Comp[,2] - min(Comp[,2]))/(max(Comp[,2])
- min(Comp[,2]))
```

  (ii) Calculate the output of the power mean for $p = -1, p = 0, p = 1$ and $p = 2$.

We provide the outputs for the first 5 instances.				
For $p = -1$				
0.2217651	0.3082053	0.4656616	0.1761840	0.3541126
For $p = 0$				
0.2963155	0.3271748	0.4705217	0.2283083	0.3542787
For $p = 1$				
0.3959274	0.3473119	0.4754325	0.2958536	0.3544450
For $p = 2$				
0.4750938	0.3663437	0.4802931	0.3506195	0.3546112

  (iii) What is the value of accuracy and time that has the best aggregated value using each of the values of p?

In all cases, the best aggregated score was obtained for the 73rd entry, which had original values
0.981   0.9822697.
The aggregated scores for each p were,

For $p = -1$	0.9442979
For $p = 0$	0.9453090
For $p = 1$	0.9463212
For $p = 2$	0.9473323

(iv) Plot the outputs for each of the functions and compare the results.

> This time we will plot the outputs for $p = -1$ against $p = 2$. We can
> see that along the line $x = y$, there are a number of values—this is
> because both power means are idempotent, so when the two inputs are
> almost the same, the output will be close to this line.
>
> Aside from this, however, we can see that the values for $p = 2$ are
> always greater than for $p = -1$ (upper half of the triangle) and that
> there are sometimes big differences for low values.

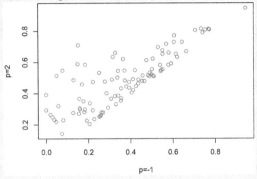

Chapter 3 Weighted Aggregation: Solutions

3.9   Practice Questions

1. There are 6 senior employees at *Lebeau Industries* who will each evaluate
   candidates for a job position. There are 3 team leaders, 2 managers (who each
   should have twice as much influence as any of the team leaders) and 1 executive
   (who should have 1.5 times as much influence as either of the managers).

   (i) What would be an appropriate weighting vector if the scores for each
        candidate are to be aggregated using a weighted power mean?
   (ii) What would you use as the value for p? Why?

> (i) To get the ratios as they are desired, we can start by giving each of the
>   team leaders an importance of 1, the two managers an importance of
>   2, and the executive an importance of 3. We can then normalize this
>   allocation so that it adds to 1.
>
> | Team leaders | | | Managers | | Executive |
> |---|---|---|---|---|---|
> | w_1 | w_2 | w_3 | w_4 | w_5 | w_6 |
> | 1 | 1 | 1 | 2 | 2 | 3 |
>
> Since the total is 10, an appropriate weighting vector would be

$$\mathbf{w} = \langle 0.1, 0.1, 0.1, 0.2, 0.2, 0.3 \rangle$$

The most important thing is that the length of the weighting vector is 6. The weights in $\mathbf{w}$ here are indicative of the preferences but anything similar would be okay depending on how literally the original ratios were interpreted.

(ii) If it is desired that most of the senior employees rate the candidates highly for a high score to be awarded, then values of p below 1 are best. Any value of p below zero will mean that if any of the 6 give a score of zero, then the candidate's score will immediately be zero. This means each of the 6 senior employees effectively have a veto option.

On the other hand, if it is desired that high scores from just a few can still result in a high candidate score, then higher values of p could be used.

In this scenario, a more consensual opinion is likely to be desired, so anything below 1 would probably be more reasonable.

2. Students applying for a foreign scholarship are to be given an overall excellence evaluation based on their performance across 4 subjects: mathematics, computer science, English and Chinese. The scholarship is in China and so Chinese should be weighted highest, however the scores in mathematics and computer science are more important than English scores.

   (i) What would be an appropriate weighting vector if the scores for each candidate are to be aggregated using a geometric mean?

   (ii) What are the advantages of using a geometric mean here rather than an arithmetic mean? What are the disadvantages?

(i) Any weighting vector that has the highest weight for Chinese and the lowest weight for English would be acceptable. Associating w_1 with mathematics, w_2 with computer science, w_3 with English and w_4 with Chinese, an example weighting vector would be

$$\mathbf{w} = \langle 0.25, 0.25, 0.2, 0.3 \rangle.$$

Whether a weighted arithmetic mean, geometric or other power mean is used, these weights still give more preference to a student who does well in Chinese.

(ii) A geometric mean has the absorbent element of 0, so that any student scoring a 0 will immediately have an overall score of 0. The output will also generally tend toward the lower inputs, meaning that all around performance across all subjects is preferred to a student that only scores very will in one subject.

3. Suppose we have three candidates for a job interview and four selection committee members give the following evaluations out of 10.

Candidate	SC 1	SC 2	SC 3	SC 4
A	8	7	9	1
B	3	9	8	5
C	6	4	8	8

(i) What would be the overall score for each candidate using the weighted arithmetic mean if the selection committee are weighted by importance with the vector $\mathbf{w} = \langle 0.3, 0.3, 0.2, 0.2 \rangle$?

(ii) What would be the overall score for each candidate using the geometric mean with the same weighting vector?

(iii) What would be the Borda count score if each first place vote is worth 10 points, each second place vote is worth 5 points, and each third place vote is worth 1 point?

(i) Candidate scores using the weighted arithmetic mean

A: $0.3(8) + 0.3(7) + 0.2(9) + 0.2(1) = 2.4 + 2.1 + 1.8 + 0.2 = 6.5$

B: $0.3(3) + 0.3(9) + 0.2(8) + 0.2(5) = 0.9 + 2.7 + 1.6 + 1 = 6.2$

C: $0.3(6) + 0.3(4) + 0.2(8) + 0.2(8) = 1.8 + 1.2 + 1.6 + 1.6 = 6.2$

(ii) Candidate scores using the weighted geometric mean

A: $8^{0.3} \times 7^{0.3} \times 9^{0.2} \times 1^{0.2} = 5.191644$

B: $3^{0.3} \times 9^{0.3} \times 8^{0.2} \times 5^{0.2} = 5.621098$

C: $6^{0.3} \times 4^{0.3} \times 8^{0.2} \times 8^{0.2} = 5.960729$

(note that Candidate C is the best performing here, since he/she has a lowest score of 4)

(iii) Using a Borda count, the scores for each candidate across the selection criteria would be:

Candidate	SC 1	SC 2	SC 3	SC 4	Total
A	10	5	10	1	26
B	1	10	3	5	19
C	5	1	3	10	19

(note that for SC3, since B and C were tied, the 1 and the 5 for second and third place were added and split between them)

The best candidate using this Borda count scoring system would be Candidate A.

4. Suppose we have three classification algorithms and we test them on four datasets. The results (as percentages) and size of each dataset are given as follows

Classifier	Set 1 (n=89)	Set 2 (n=27)	Set 3 (n=161)	Set 4 (n=68)
Neural network	71	20	87	56
Decision Tree	73	29	94	51
knn	81	64	95	49

(i) Use a weighted arithmetic mean (and justify your choice of weighting vector) to give an overall evaluation of each of the classifiers.

(ii) Using the same weighting vector, use a quadratic mean (a power mean with $p = 2$) to evaluate the methods. Is the relative ranking the same.

(iii) Use a Borda count-based rule in order to rank the different classifiers.

(i) There are a few justifiable approaches to deciding on the weights here. One way would be to base it on the number of data, in which case we would have something like

$$\mathbf{w} = \left\langle \frac{89}{345}, \frac{27}{345}, \frac{161}{345}, \frac{68}{345} \right\rangle$$

$$= \langle 0.258, 0.078, 0.467, 0.197 \rangle,$$

however we could also note that it seems harder to do well on set 2 and 4, with very low percentages for these two datasets. We could then instead use a weighting vector like

$$\mathbf{w} = \langle 0.2, 0.4, 0.1, 0.3 \rangle$$

which gives more weight to the 'harder' sets (and of course we could also use an unweighted arithmetic mean).

Using $\mathbf{w} = \langle 0.2, 0.4, 0.1, 0.3 \rangle$, the resulting overall scores would be:

Neural network: $0.2(71) + 0.4(20) + 0.1(87) + 0.3(56) = 47.7$

Decision Tree: $0.2(73) + 0.4(29) + 0.1(94) + 0.3(51) = 50.9$

knn: $0.2(81) + 0.4(64) + 0.1(95) + 0.3(49) = 66.$

(ii) With a quadratic mean, these would become:

Neural network: $\left(0.2(71^2) + 0.4(20^2) + 0.1(87^2) + 0.3(56^2) \right)^{1/2} = 53.5341$

Decision Tree: $\left(0.2(73^2) + 0.4(29^2) + 0.1(94^2) + 0.3(51^2) \right)^{1/2} = 55.3724$

knn: $\left(0.2(81^2) + 0.4(64^2) + 0.1(95^2) + 0.3(49^2) \right)^{1/2} = 67.6269.$

With these weights, the ranking doesn't change.

(iii) Using a standard Borda count, the scores would be as follows,

Candidate	Set 1	Set 2	Set 3	Set 4	Overall Score
Neural network	0	0	0	2	2
Decision tree	1	1	1	1	4
knn	2	2	2	0	6

As with the previous results, the ranking in terms of best performing classifier is knn, then Decision tree, then Neural network.

---

5. The rankings of a set of schools was released with the claim that 15 % of the outcome was based on the school environment, 25 % was based on student performance and 60 % was based on student progress. It was argued that a weighted median should be used to calculate each school's overall score (because medians are more robust to outliers). Do you agree?

> Medians being robust to outliers is useful when we're aggregating a lot of data relating to the same variable, however in this scoring system, an 'outlier' would be interpreted as a value for a single school that was extremely different to its other two scores, e.g. if they scored 10/10 for school environment, 9.8/10 for student performance and 0/10 for student progress, then the last 0 could perhaps be called an outlier. So even a standard median doesn't really make sense in this case.
>
> More importantly, a weighted median with this vector would always give an output equal to the student progress score (since it has more than 50 % of the weight). In this case, it would effectively ignore the scores in the other categories. The argument does not make much sense.

6. A weighted power mean is used (with $p = 1/5$) to evaluate different land management plans in terms of biodiversity conservation. The criteria are 1. positive impact to rare species ($w_1 = 0.3$), 2. positive impact to native species ($w_2 = 0.2$), 3. conservation of native vegetation ($w_3 = 0.25$), 4. economic benefit to local (human) population ($w_4 = 0.1$), and 5. cost ($w_5 = 0.15$).

   (i) Calculate the score for a management plan with partial evaluations denoted by $\mathbf{x} = \langle 0.6, 0.5, 0.9, 0.7, 0.3 \rangle$.
   (ii) If the plan could be slightly improved with respect to one of the criteria, which would you choose?

> (i) The overall score for this plan would be
> $(0.3(0.6^{0.2}) + 0.2(0.5^{0.2}) + 0.25(0.9^{0.2}) + 0.1(0.7^{0.2}) + 0.15(0.3^{0.2}))^5 \approx$ 0.593.
> (ii) The value of $p = 1/5$ means the output tends toward lower values. So it is likely that it will be improved by either improving the score for cost, or the score for impact to native species, however the highest weight is allocated to impact to rare species.
>
> By comparing the effects of increasing each score by 0.01, it turns out that increasing the first input is the best option (impact to rare species).

3.11   Practice Questions Using R

1. Suppose you have an input vector $\mathbf{x} = \langle 0.88, 0.12, 0.06, 0.46, 0.11 \rangle$ and a weighting vector $\mathbf{w} = \langle 0.26, 0.21, 0.14, 0.39, 0 \rangle$, calculate the weighted power mean for the following cases

   (i) $p = 1$
   (ii) $p = 2$
   (iii) $p = 0$
   (iv) $p = -1$

   and comment on (i.e. compare) the results.

> (i) 0.4418 (ii) 0.5360933 (iii) 0.3087524 (iv) 0.1913285.
> Lower values of p make the output tend toward the lower inputs however it is always between the minimum and maximum (0.06 and 0.88). The fourth input has the highest weight allocated, so the overall output when p is closer to 1 shouldn't be too far away from this.

2. Suppose you have the input vector $\mathbf{x} = \langle 0.64, 0.50, 0.35, 0.93, 0 \rangle$ and weighting vector $\mathbf{w} = \langle 0.15, 0.34, 0.16, 0.02, 0.33 \rangle$, calculate the weighted power mean for the following cases

   (i) $p = 1.5$
   (ii) $p = 2.5$
   (iii) $p = 0$
   (iv) $p = -1.5$

   and comment on (i.e. compare) the results.

> (i) 0.3985532 (ii) 0.45455 (iii) 0 (iv) 0.
> Since there is an input equal to 0, the output becomes zero for any values of p equal to zero or less. For $p > 1$, the output is closer to the high inputs.

3. For $\mathbf{x} = \langle 0.37, 0.97, 0.01, 0.84, 0.03 \rangle$ and $\mathbf{w} = \langle 0.24, 0.06, 0.28, 0.14, 0.28 \rangle$, calculate the upper and lower weighted medians.

> Both the upper and lower weighted medians are equal to 0.03. This is because the combined weight allocated to the two lowest inputs is 0.56 (greater than 50 %).

4. Use the female students volleyball dataset and determine the best two players overall using a Borda count type approach where being ranked first in any category is worth 5 points, second in any category is worth 3 points, and third in any category is worth 1 point (zero points for any lower than 3rd).

> Given the volley matrix V, the data can be transformed using
> ```
> V[,j] <- rank(V[,j])
> ```
> for each row. This will mean that quicker times will have a lower (number) rank allocated, while for the remaining values a higher rank score means better performance. One thing to keep in mind with the rank function is that if scores are tied, the ranks will be divided up, so we will have half-ranks. For the sprint times, we want ranks of 1 to be allocated 5 points, ranks of 2 to have 3 points, and ranks of 3 to be given 1 point. We can give half ranks of 1.5 a score of 4, half ranks of 2.5 a score of 2 and everything else zero. If there are multiple girls tied for third place (they will have a score of 17.5 or 3.5 depending on whether higher or lower is better), we will give a score of zero.

The following can be used to make everything above 3 in the sprint ranks a 0.

```
V[V[,1]>3.5,1] <- 0.
```

The scoring can then be allocated as follows.

```
V[,1] <- c(0,1,2,3,4,5)[match(V[,1],c(0,3,2.5,2,1.5,1))]
```

For the remaining variables we will need to use

```
V[V[,j]<18,j] <- 0.
V[,j] <- c(0,1,2,3,4,5)[match(V[,j],c(0,18,18.5,19,19.5,20))]
```

Then to sum the scores, we can do the following

```
totals <- array(0,20)
for(i in 1:20) totals[i] <- sum(V[i,])
```

To find the best scoring girls, we can use which.max() or order()

```
order(totals,decreasing = TRUE)
```

The first two values returned are 8 and 19, corresponding with Kayoko and Chisato. Chisato has the best endurance score and the second best sprint time, while Kayoko is the tallest and has the second best endurance score (8 points each).

Chapter 4 Averaging with Interaction: Solutions

4.6   Practice Questions

1. Calculate the outputs for an OWA operator with $\mathbf{w} = \langle 0.1, 0.4, 0.3, 0.2 \rangle$ and

   (i) $\mathbf{x} = \langle 0.8, 0.2, 1, 1 \rangle$
   (ii) $\mathbf{x} = \langle 0.9, 0.1, 0.7, 0.6 \rangle$

$$(i)\ OWA(\mathbf{x}) = 0.1(0.2) + 0.4(0.8) + 0.3(1) + 0.2(1)$$
$$= 0.02 + 0.32 + 0.3 + 0.2 = 0.84$$
$$(ii)\ OWA(\mathbf{x}) = 0.1(0.1) + 0.4(0.6) + 0.3(0.7) + 0.2(0.9)$$
$$= 0.01 + 0.24 + 0.21 + 0.18 = 0.64$$

2. Calculate the outputs for an OWA operator with $\mathbf{w} = \langle 0.1, 0, 0.1, 0.5, 0.3 \rangle$ and

   (i) $\mathbf{x} = \langle 0.9, 0.3, 0.5, 0.8, 0 \rangle$
   (ii) $\mathbf{x} = \langle 0.2, 0.1, 0.1, 0.9, 1 \rangle$

(i) $\text{OWA}(\mathbf{x}) = 0.1(0) + 0 + 0.1(0.5) + 0.5(0.8) + 0.3(0.9)$

$$= 0.05 + 0.4 + 0.27 = 0.72$$

(ii) $\text{OWA}(\mathbf{x}) = 0.1(0.1) + 0 + 0.1(0.2) + 0.5(0.9) + 0.3(1)$

$$= 0.01 + 0.02 + 0.45 + 0.3 = 0.78$$

3. Calculate the output for a discrete Choquet integral for $\mathbf{x} = \langle 0.8, 0.3, 0.1 \rangle$ when the associated fuzzy measure is given by

$$v(\{1, 2, 3\}) = 1$$

$$v(\{1, 2\}) = 0.9 \quad v(\{1, 3\}) = 0.2 \quad v(\{2, 3\}) = 0.3$$

$$v(\{1\}) = 0.2 \qquad v(\{2\}) = 0.3 \qquad v(\{3\}) = 0.1$$

$$v(\emptyset) = 0$$

The ordering (from lowest to highest) is x_3, x_2, x_1, so we will have (writing the v sets in abbreviated form)

$\text{Ch}_v = 0.1(v(123) - v(12)) + 0.3(v(12) - v(1)) + 0.8(v(1) - v(\emptyset))$

$$= 0.1(1 - 0.9) + 0.3(0.9 - 0.2) + 0.8(0.2)$$

$$= 0.1(0.1) + 0.3(0.7) + 0.8(0.2) = 0.01 + 0.21 + 0.16 = 0.38$$

4. Calculate the output for a discrete Choquet integral for $\mathbf{x} = \langle 1, 8, 12, 7 \rangle$ when the associated fuzzy measure is given by

$$v(\{1, 2, 3, 4\}) = 1$$

$$v(\{1, 2, 3\}) = 0.9 \quad v(\{1, 2, 4\}) = 0.9 \quad v(\{1, 3, 4\}) = 0.9 \quad v(\{2, 3, 4\}) = 0.9$$

$$v(\{1, 2\}) = 0.2 \quad v(\{1, 3\}) = 0.2 \quad v(\{1, 4\}) = 0.2$$

$$v(\{2,3\}) = 0.9 \quad v(\{2,4\}) = 0.2 \quad v(\{3,4\}) = 0.3$$

$$v(\{1\}) = 0.2 \quad v(\{2\}) = 0.3 \quad v(\{3\}) = 0.1 \quad v(\{4\}) = 0.1$$

$$v(\emptyset) = 0$$

The ordering (from lowest to highest) is x_1, x_4, x_2, x_3, so we will have

$$
\begin{aligned}
\text{Ch}_v &= 1(v(1234) - v(234)) + 7(v(234) - v(23)) + 8(v(23) - v(3)) + 12(v(3) - v(\emptyset)) \\
&= 1(1 - 0.9) + 7(0.9 - 0.9) + 8(0.9 - 0.1) + 12(0.1 - 0) \\
&= 1(0.1) + 7(0) + 8(0.8) + 12(0.1) = 0.1 + 6.4 + 1.2 = 7.7
\end{aligned}
$$

5. Define the weighting vectors for $n = 3$ and $n = 5$ using the quantifier $Q(t) = t^2$.

For $n = 3$,

$$w_1 = Q\left(\frac{1}{3}\right) - Q(0) = \left(\frac{1}{3}\right)^2 = \frac{1}{9}$$

$$w_2 = Q\left(\frac{2}{3}\right) - Q\left(\frac{1}{3}\right) = \left(\frac{2}{3}\right)^2 - \left(\frac{1}{3}\right)^2 = \frac{4}{9} - \frac{1}{9} = \frac{3}{9} = \frac{1}{3}$$

$$w_2 = Q\left(\frac{3}{3}\right) - Q\left(\frac{2}{3}\right) = (1)^2 - \left(\frac{2}{3}\right)^2 = 1 - \frac{4}{9} = \frac{5}{9}$$

hence $\mathbf{w} = \left\langle \frac{1}{9}, \frac{1}{3}, \frac{5}{9} \right\rangle$.
  For $n = 5$,

$$w_1 = Q\left(\frac{1}{5}\right) - Q(0) = \left(\frac{1}{5}\right)^2 = \frac{1}{25}$$

$$w_2 = Q\left(\frac{2}{5}\right) - Q\left(\frac{1}{5}\right) = \left(\frac{2}{5}\right)^2 - \left(\frac{1}{5}\right)^2 = \frac{4}{25} - \frac{1}{25} = \frac{3}{25}$$

$$w_3 = Q\left(\frac{3}{5}\right) - Q\left(\frac{2}{5}\right) = \left(\frac{3}{5}\right)^2 - \left(\frac{2}{5}\right)^2 = \frac{9}{25} - \frac{4}{25} = \frac{5}{25} = \frac{1}{5}$$

$$w_4 = Q\left(\frac{4}{5}\right) - Q\left(\frac{3}{5}\right) = \left(\frac{4}{5}\right)^2 - \left(\frac{3}{5}\right)^2 = \frac{16}{25} - \frac{9}{25} = \frac{7}{25}$$

$$w_5 = Q\left(\frac{5}{5}\right) - Q\left(\frac{4}{5}\right) = (1)^2 - \left(\frac{4}{5}\right)^2 = 1 - \frac{16}{25} = \frac{9}{25}$$

hence $\mathbf{w} = \langle \frac{1}{25}, \frac{3}{25}, \frac{1}{5}, \frac{7}{25}, \frac{9}{25} \rangle$.

6. Compare the weighting vectors for $Q(t) = \sqrt{t}$ (Example 4.5) and $Q(t) = t^2$ in the previous question. For quantifiers of the form $Q(t) = t^q$, what can you predict about the orness of the weighting vectors when $q > 1$ or $q < 1$?

For t^2, the weighting vectors make an increasing sequence, which means that these will tend more toward the higher inputs (and hence high orness). On the other hand, the sequence for $\sqrt{t}$

$$\mathbf{w} = \langle 0.4472, 0.1852, 0.1421, 0.1198, 0.1056 \rangle$$

is decreasing, which means that more weight is allocated to the lower inputs and hence the function will have a lower orness. For $q < 1$ the orness will be low, for $q > 1$, the orness will be high. (this corresponds with the parameter p in power means). However note that for quantifiers, we can't have negative values of q since we need the function to be increasing.

7. Define a weighting vector for $n = 7$ that can approximate the natural language quantifier "at least 3".

Here we are thinking of statements like "at least 3 of the inputs are high" or "at least 3 of the criteria are satisfied". So if we have an input vector $\mathbf{x} = \langle 0, 0, 0, 0, 0, 1, 1 \rangle$ then this probably should not satisfy the statement. A weighting vector $\mathbf{w} = \langle 0, 0, 0, 0, 1, 0, 0 \rangle$ allocates all weight to the third highest input. Let's take a look at a few input vectors and their output.

Input	Output
$\langle 0, 0, 1, 0, 1, 0, 1 \rangle$	1
$\langle 0, 0, 1, 0, 0.5, 0, 1 \rangle$	0.5
$\langle 0, 0, 0.3, 0, 0.5, 0, 1 \rangle$	0.3
$\langle 1, 0.3, 0.8, 0.1, 0.4, 0.3, 0.2 \rangle$	0.4

So essentially, the output indicates that at least three of the inputs are this high (or higher).

8. Define a weighting vector for $n = 4$ that can approximate the natural language quantifier "almost all".

The quantifier for "all" is equivalent to the minimum function (so with $\mathbf{w} = \langle 1, 0, \ldots, 0, 0 \rangle$). For *almost* all, we can focus either on the second lowest or two lowest inputs. A weighting vector $\mathbf{w} = \langle 0.4, 0.6, 0, 0 \rangle$ would spread the weight between the two lowest inputs. So if we had $\mathbf{x} = \langle 0, 1, 1, 1 \rangle$ the output would be 0.6, however $\mathbf{x} = \langle 0, 0.5, 1, 1 \rangle$ would only have an output of 0.3.

9. The manager of a clothing store forecasts her monthly sales based on an OWA of the previous 4 months. Interpret the vectors $\mathbf{w} = \langle 0.1, 0.2, 0.3, 0.4 \rangle$ and $\mathbf{w} = \langle 0.4, 0.3, 0.2, 0.1 \rangle$ in terms of whether they give a 'conservative' or 'optimistic' estimate of the next month's sales.

The vector $\mathbf{w} = \langle 0.1, 0.2, 0.3, 0.4 \rangle$ is weighted toward higher inputs. It means that she is basing her forecasts predominantly on the best months of sales. This means the prediction will be optimistic. On the other hand, for $\mathbf{w} = \langle 0.4, 0.3, 0.2, 0.1 \rangle$, the weight is mainly based on the months where sales were low, so we can say this is conservative.

10. What would be the fuzzy measure used to define a Choquet integral that is equivalent to an OWA function with weights $\mathbf{w} = \langle 0.1, 0.7, 0.2 \rangle$?

Recall that a Choquet integral is equivalent to an OWA if the measure of singletons are the same, the measure of pairs are the same, measure of triples are the same, etc. Also recall that in calculation of a Choquet integral, the weight allocated to the highest input is the fuzzy measure of a singleton, while the weight associated with the lowest input is the weight of the set excluding the lowest input subtracted from 1. The fuzzy measure will be

$$v(\{1, 2, 3\}) = 1$$

$$v(\{1, 2\}) = v(\{1, 3\}) = v(\{2, 3\}) = 0.9$$

$$v(\{1\}) = v(\{2\}) = v(\{3\}) = 0.2$$

$$v(\emptyset) = 0$$

i.e. because now $v(\{1, 2, 3\}) - v(\{1, 2\}) = 0.1$ (which would be applied to the lowest weight, and is the same for any pair) and the weight applied to the middle weight would be $v(\{1, 2\}) - v(\{1\}) = 0.7$ (or any singleton subtracted from any pair) and the weight allocated to the largest input is 0.2.

11. What would be the fuzzy measure used to define a Choquet integral equivalent to a WAM with $\mathbf{w} = \langle 0.3, 0.45, 0.25 \rangle$?

> Here we would have $v(\{1\}) = 0.3, v(\{2\}) = 0.45, v(\{3\}) = 0.25$ and the pairs and triples can be found by adding these together. So,
> $v(\{1, 2\}) = 0.3 + 0.45 = 0.75,$
> $v(\{1, 3\}) = 0.3 + 0.25 = 0.55,$
> $v(\{2, 3\}) = 0.45 + 0.25 = 0.7,$
> $v(\{1, 2, 3\}) = 0.3 + 0.45 + 0.25 = 1$

12. Will the fuzzy measure below define a Choquet integral that is equivalent to either the weighted arithmetic mean or the OWA? Explain why/why not.

$$v(\{1, 2, 3\}) = 1$$

$$v(\{1, 2\}) = 0.7 \quad v(\{1, 3\}) = 0.6 \quad v(\{2, 3\}) = 0.8$$

$$v(\{1\}) = 0.2 \quad v(\{2\}) = 0.5 \quad v(\{3\}) = 0.4$$

$$v(\emptyset) = 0$$

> By looking at the fuzzy measure we see that the values of the singletons are all different, so it can't be the same as an OWA. We also note that the measure of $\{2, 3\}$ is not equal to $v(\{2\}) + v(\{3\}) = 0.5 + 0.4 = 0.9$. So in this case, the fuzzy measure is **not** equivalent to either a WAM or an OWA.

13. Will the fuzzy measure below define a Choquet integral that is equivalent to either the weighted arithmetic mean or the OWA? Explain why/why not.

$$v(\{1, 2, 3\}) = 1$$

$$v(\{1, 2\}) = 0.7 \quad v(\{1, 3\}) = 0.7 \quad v(\{2, 3\}) = 0.7$$

$$v(\{1\}) = 0 \quad v(\{2\}) = 0 \quad v(\{3\}) = 0$$

$$v(\emptyset) = 0$$

> Since the values are the same when the size of the sets are the same, this must be equivalent to an OWA. In particular, it is equivalent to an OWA with $\mathbf{w} = \langle 0.3, 0.7, 0 \rangle$.

14. Suppose we have four job applicants who want a job on our data analysis team. We need them to be *either* strong in both coding and mathematics, *or* strong communicators. They are given scores out of 10 for each of the criteria.

Candidate	JA 1	JA 2	JA 3	JA 4
Coding	9	3	2	7
Mathematics	2	8	3	8
Communication	4	6	8	3

(i) Define a fuzzy measure that can model our requirements. [Hint: the value assigned to *coding* and *mathematics* should be high but the values assigned to these two criteria individually should be very low. The value assigned to *communication* on its own should also be high.]

(ii) Evaluate the 4 candidates using the Choquet integral with respect to the fuzzy measure you defined in (i).

(i) For coding = 1, mathematics = 2 and communication = 3, we can assign $v(\{1,2\}) = 1, v(\{3\}) = 1$ then everything else can be determined from monotonicity requirements from the fuzzy measure (and we can set $v(\{1\}) = v(\{2\}) = 0$ so that we need both of these to be high. The resulting fuzzy measure is

$$v(\{1,2,3\}) = 1$$

$$v(\{1,2\}) = 1 \quad v(\{1,3\}) = 1 \quad v(\{2,3\}) = 1$$

$$v(\{1\}) = 0 \quad v(\{2\}) = 0 \quad v(\{3\}) = 1$$

$$v(\emptyset) = 0$$

Note that $v(\{1,3\})$ and $v(\{2,3\})$ are both equal to 1 as well since they include 3.

(ii) (using abbreviated notation)
For JA 1, the output will be

$$2(v(123)-v(13))+4(v(13)-v(1))+9(v(1)-0) = 2(0)+4(1)+9(0) = 4$$

For JA 2,

$$3(v(123)-v(23))+6(v(23)-v(2))+8(v(2)-0) = 2(0)+6(1)+9(0) = 6$$

For JA 3,

$$2(v(123)-v(23))+3(v(23)-v(3))+8(v(3)-0) = 2(0)+3(0)+8(1) = 8$$

For JA 4,

$$3(v(123)-v(12))+7(v(12)-v(2))+8(v(2)-0) = 3(0)+7(1)+8(0) = 7$$

So job applicant 3's score is better than 4's because the communication score is higher than the minimum coding/maths score for JA 4.

15. For the fuzzy measure below, which of the criteria seems most important overall? Which seems least important?

$$v(\{1, 2, 3\}) = 1$$

$$v(\{1, 2\}) = 0.8 \quad v(\{1, 3\}) = 0.9 \quad v(\{2, 3\}) = 0.4$$

$$v(\{1\}) = 0.1 \quad v(\{2\}) = 0.2 \quad v(\{3\}) = 0.4$$

$$v(\emptyset) = 0$$

Looking at the singletons, the value for $v(\{3\})$ is highest, so in isolation, the third criterion seems the most important. However looking at the first criterion, when it's by itself it might not be worth much but it has very high values when with both 2 and 3. The Shapley index for the first criterion is

$$\phi(1) = \frac{(3-2-1)!2!}{3!}(v(123) - v(23)) + \frac{(3-1-1)!1!}{3!}(v(12) - v(2))$$

$$+ \frac{(3-1-1)!1!}{3!}(v(13) - v(3)) + \frac{(3-0-1)!0!}{3!}(v(1) - v(\emptyset))$$

$$= \frac{2}{6}(1 - 0.4) + \frac{1}{6}(0.8 - 0.2) + \frac{1}{6}(0.9 - 0.4) + \frac{2}{6}(0.1))$$

$$= \frac{1.2 + 0.6 + 0.5 + 0.2}{6} = 0.41667$$

On the other hand, for criterion 3,

$$\phi(3) = \frac{(3-2-1)!2!}{3!}(v(123) - v(12)) + \frac{(3-1-1)!1!}{3!}(v(23) - v(2))$$

$$+ \frac{(3-1-1)!1!}{3!}(v(13) - v(1)) + \frac{(3-0-1)!0!}{3!}(v(3) - v(\emptyset))$$

$$= \frac{2}{6}(1 - 0.8) + \frac{1}{6}(0.4 - 0.2) + \frac{1}{6}(0.9 - 0.1) + \frac{2}{6}(0.4))$$

$$= \frac{0.4 + 0.2 + 0.8 + 0.8}{6} = 0.3667$$

This confirms that, on average, the first criterion is more important.

4.7 Practice Questions Using R

1. Use `rand.x <- sample(1:1000,10)` to create a random input vector and
   calculate the OWA with weights defined by a quantifier using
   `w <- ((1:10)/10)^2 - ((0:9)/10)^2.`

   > The above calculation (for the weighting vector) gives,
   > 0.01 0.03 0.05 0.07 0.09 0.11 0.13 0.15 0.17 0.19
   > Creating a random input vector, we obtain (for example),
   > 365   355   396   98   586   100   699   50   563   35
   > The calculation for this vector will be 454.21.

2. Use the method from the previous question to compare the outputs of OWA
   functions and power means. For power means with power p and quantifier-based
   OWAs that use $Q(t) = t^q$, look at whether or not the functions tend toward higher
   or lower inputs. The quantifier in the previous question is based on $Q(t) = t^2$ (for
   $n = 10$), changing the 2 to a 3 will mean it is uses $Q(t) = t^3$ and so on.

   > Using the same vector as above, here is how the output values will change
   > with p (for unweighted power means) and q for quantifier-based OWAs.
   >
   > | p | PM | q | OWA |
   > | --- | --- | --- | --- |
   > | −5 | 53.67273 | 0.2 | 116.0364 |
   > | −2 | 83.03456 | 0.5 | 211.8399 |
   > | 0.5 | 272.6015 | 1 | 324.7 |
   > | 1 | 324.7 | 2 | 454.21 |
   > | 2 | 398.5425 | 3 | 522.847 |
   > | 5 | 500.6106 | 5 | 591.3157 |
   >
   > So we can see as q is closer to zero, the output for the OWA decreases. At the
   > limit, it would be equivalent to the minimum function. For $p = 1$ and $q = 1$,
   > both types of operators become equivalent to the arithmetic mean and so we
   > obtain the standard average. Then increasing q for quantifier-based OWAs
   > has the same effect as increasing p for unweighted power means. For q
   > approaching infinity, we will obtain a weighting vector that makes the OWA
   > equivalent to the maximum.

3. Open the data file `FMEASURES.txt` and assign the two columns as vectors to
   fm.1 and fm.2 (or any assignment you like—these are fuzzy measures for five
   variables).

   (i) Calculate the output for $\mathbf{x} = \langle 5, 15, 13, 8, 9 \rangle$ using fm.1 and fm.2.
   (ii) Calculate the Shapley values
   (iii) Calculate the orness values.
   (iv) Explain why fm.1 or fm.2 resulted in a higher output with reference to the
   Shapley and orness calculations.

Once the file is in the working directory, it can be opened using
```
FMS <- as.matrix(read.table("FMEASURES.txt"))
```
Then we can use `fm.1 <- FMS[,1]` and `fm.2 <- FMS[,2]` to set the fuzzy measures as vectors.

(i) For `fm.1`, the output should be 10.77273, for `fm.2`, it should be 10.14197.

(ii) Using the `shapley()` function from the R tutorial, for `fm.1`, the shapley values are
```
0.1649531 0.1356537 0.2484739 0.1292385 0.3216809
```
(i.e. the most important inputs on average are the third and fifth).
For `fm.2`,
```
0.1561213 0.2428396 0.1523530 0.1972086 0.2514775.
```

(iii) The orness calculations respectively are 0.4925351 and 0.5234101, which means both fuzzy measures are close to treating high and low inputs equally on average.

(iv) The differences are only slight. From the Shapley values, we might have expected the second fuzzy measure to have produced a larger value since it allocates more weight to the highest input and less weight to the lowest input. However we can see with the inputs equal to 8 and 9, the first fuzzy measure allocates 3 times as much weight to the 9 than to the 8, while for `fm.2` the weights are more similar.

We also might have expected the second fuzzy measure to produce a higher value based on the orness, which means it tends slightly more towards high values than `fm.1`.

However remember that Shapley indices and the orness calculation are only what happens on average. Specific vectors (and their orderings) will result in specific weights. In particular, here the weights applied to each input will be

	$v(12345) - v(2345)$	$v(2)$	$v(23)\text{-}v(3)$	$v(2345) - v(235)$	$v(235) - v(23)$
	x_1	x_2	x_3	x_4	x_5
fm.1	0		0.02273	0.40909	0	0.56818
fm.2	0		0.03253	0.23671	0	0.73077

and so we can see that the first fuzzy measure just happens to allocate more weight to the third input in this case.

Chapter 5 Fitting Aggregation Functions to Empirical Data: Solutions

5.6   Practice Questions Using R

1. The Kei Hotels rating data is a 56×10 table where the first column indicates Kei's ratings for each hotel (out of 100) and columns 2 to 9 are the ratings of similar users.

   (i) Download the `KeiHotels.txt` file and save it to your R working directory.
   (ii) Assign the data to a matrix, e.g. using
        ```
        kei.data <- as.matrix(read.table("KeiHotels.txt"))
        ```
   (iii) Define a function to measure the similarity between Kei and the other online users. (The Euclidean distance can be defined using the Minkowski distance with $p = 2$)
   (iv) Which of the users is *most similar* to Kei? Investigate using scatterplots, histograms, the correlation and similarity between Kei and the other users.

The Minkowski distance (with Manhattan distance as the default) can be defined using
```
minkowski <- function(x,y,p=1) (sum(abs(x-y)^p))^(1/p)
```
Using $p = 2$, i.e.
```
minkowski(kei.data[,1],kei.data[,2],2)
```
gives the Euclidean distance between Kei and user 2, while
```
minkowski(kei.data[,1],kei.data[,3])
```
gives the Manhattan distance between Kei and user 3.
We can also use correlation (which is one of the standard statistical functions in R).
```
cor(kei.data[,1],kei.data[,3])
```
gives the Pearson correlation, while
```
cor(kei.data[,1],kei.data[,3], method = "spearman")
```
gives Spearman correlation.
To calculate all values at once (rather than one at a time), we can use
```
d2 <- array(0,9)
for(i in 1:9) {
d2[i] <- minkowski(kei.data[,1],kei.data[,i+1],2) }
```
We will obtain the following results.

Sim	2	3	4	5	6	7	8	9	10
Euclidean	103.407	119.633	116.837	**73.239**	75.478	127.902	119.105	94.557	107.476
Manhattan	541	564	677	**358**	385	691	552	481	551
Pearson	0.879	0.865	0.853	**0.947**	0.941	0.855	0.851	0.921	0.886
Spearman	0.879	0.865	0.853	**0.947**	0.941	0.855	0.851	0.921	0.886

Remember that the lower the distances, the more similar, while the correlation should be higher for higher similarity. So the most similar user from these indicators is user 5. However there are other ways that we could consider similarity.

We can compare distributions using histograms. These do not tell us about which hotels were rated, however we can spot whether users tend to rate hotels as consistently high, consistently low, etc. Below are histograms of User 5 (left), Kei (centre) and user 7 (right). Even though User 5 is the most similar user in terms of the other measures, a rough look at these histograms suggests that user 7 may have a more similar pattern in terms of tendency to rate highly or lowly.

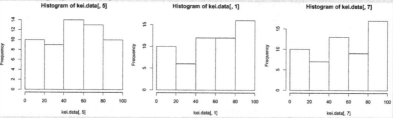

Scatterplots will most likely reflect the values obtained using correlation, however they can also help us pick up whether there could be a non-linear relationship (which would not be reflected in the correlation values). The scatterplots below plot Kei's scores against user 2 (left) and user 5 (right).

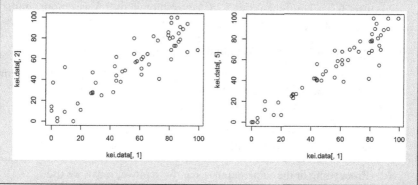

2. Download the AggWaFit R file to your working directory and load into the R workspace using,

```
source("AggWaFit718.R")
```

   (i) Using fit.QAM, find the weights for a weighted arithmetic mean that best approximates Kei's ratings from those of the other users.

[hint: You will need to set Kei's data as the last column. You can do this using (e.g. if your data matrix is 'A'), `A <- A[,c(2:9,1)]`

> Using
> `fit.QAM(kei.data[,c(2:10,1)])`
> the output stat file gives:
> RMSE 4.20993003328861
> Av. abs error 3.12765137378224
> Pearson correlation 0.98951986803197
> Spearman correlation 0.983894812270749
> i w-i
> 1 0.0194056285221035
> 2 0.0390876342948007
> 3 0.102891478464021
> 4 0.240427752211895
> 5 0.281991744682169
> 6 0.0470443873526975
> 7 0
> 8 0.110916927376011
> 9 0.158234447096302
> Note that the outputs of this file are 1 off the user identifiers, i.e. what we have been referring to as user 2 corresponds with w_1, user 3 with w_2 etc. Interestingly, fitting in this way the fitted weights allocate more to user 6 (w_5 here) than user 5 (w_4) even though user 5 was more similar to Kei in all of our previous investigations.

(ii) Use `fit.QAM`, `fit.OWA` to find the best weights for

- Weighted power means with $p = 0.5$, and $p = 2$, (the generators required are PM05, invPM05 and QM, invQM).
- A geometric mean (the generators are GMa and invGMa).
- An OWA.

You can also experiment with using only a subset of the variables.

> Proceeding in the same manner, e.g. for a power mean we use
> `fit.QAM(kei.data[,c(2:10,1)],g=PM05,g.inv = invPM05)`
> The RMSE and weights are shown for each of the functions below (all values to 3 decimal places)

	PM ($p = 0.5$)	PM ($p = 2$)	GM	OWA
RMSE	4.599	5.375	9.878	3.256
w_1	0.061	0.008	0.000	0.004
w_2	0.000	0.102	0.000	0.000
w_3	0.113	0.034	0.000	0.000
w_4	0.307	0.274	0.000	0.000
w_5	0.337	0.244	1.000	0.945
w_6	0.031	0.054	0.000	0.026
w_7	0.000	0.000	0.000	0.000
w_8	0.121	0.120	0.000	0.000
w_9	0.031	0.163	0.000	0.024

(in some cases the zeros here represent 0, however (especially for the geometric mean) some values are just very low.

(iii) Which model fits the data the best?

> The OWA has the lowest RMSE and so it seems to be the best fitting function. Remember that the OWA weights are not associated with users but relative values. Here the OWA is almost exactly the same as the median, with 0.945 allocated to the middle weight (whereas the median would have 1 here). The weighted arithmetic mean obtained earlier is better than any of the power means tried here—suggesting that it's better when the output doesn't tend toward high or low values (which is also supported by the fact that the OWA obtained is almost the same as the median).

(iv) Comment on similarities and differences between the users that were found to be the most similar to Kei and whether they had the highest weights allocated in the fitted data models.

> Users 5 and 6 (weights w_4 and w_5 respectively) were allocated the most weight by most of the models. Interestingly, user 5 was not always allocated the most weight and user 7 (w_6) was never allocated the least weight (even though the distance measures make it much higher).

3. Use a subset of any four of the similar users and the fit.choquet function to find the fuzzy measure that fits the data best and compare with your previous findings (trying to use more users will result in a very long time to find the values).

> Using the best 4 users based on the WAM (i.e. similar user 6,5,10,9 (in that order), the following stat file is obtained.
> RMSE 3.55879402102922
> Av. abs error 2.53337971552437
> Pearson Correlation 0.992462190684642

```
Spearman Correlation 0.98957099698045
Orness 0.529772542272563
i Shapley i
1 0.273870573870577
2 0.301551226551394
3 0.220537795537936
4 0.204040404040449
binary number nam.weights
1 0.179487179487191
2 0.214285714285701
3 0.400000000000108
4 0.117216117216137
5 0.179487179487191
6 0.954545454545463
7 0.954545454545558
8 0.0357142857143169
9 0.892857142856606
10 0.214285714285695
11 1.00000000000009
12 0.142857142857174
13 1
14 1.00000000000027
15 1.00000000000036
```

Based on the Shapley values, now the most important user is user 5 (then 6 then 10 then 9). Looking at the fuzzy measure, which has an orness suggesting it is fairly neutral (and possibly close to a weighted arithmetic mean), we can spot a few interesting relationships. $v(\{2\}|) = 0.214$ and $v(\{3\}) = 0.117$ (binary numbers $2 = 10$ and $4 = 100$ respectively), however together their weight is $v(\{2, 3\}) = 0.955$ (binary number $6 = 1010$). This suggests a complementary relationship. Similarly, 4 by itself is $v(\{4\}) = 0.036$ however with the first variable $v(\{1, 4\}) = 0.89$ (binary number $9 = 1001$). Meanwhile the variables 4 and three together are fairly additive. The superadditive/positive synergy relationship for some of these pairs simply suggests that if BOTH are high then it is likely that the output should be high, whereas in some cases, perhaps where users are more similar to each other, both having good scores does not suggest much about Kei's scores.

Index

A
aggregation function, 18, 21–23
    averaging, 21
        definition, 23
    definition, 22
    dual, 56
    fitting, 129, 143
    properties, 12
    uses of, 23
aggregation function: definition, 22
arithmetic mean, 10
    definition, 10, 11
    properties, 12
    weighted arithmetic mean, 80
array, 41

B
Borda count, 87

C
capacity, *see* fuzzy measure
Choquet integral (discrete), 108
    calculation, 111
    definition, 110
    fitting, 145
    properties, 115
    special cases, 111

D
distance
    definition, 131

E
entropy, 86

F
function, 4
    multivariate functions, 7
fuzzy measure, 108
    definition, 108

G
generalized mean, *see* power mean
geometric mean, 16
    definition, 17
    properties, 18
group decision making, 76

H
harmonic mean, 19
    definition, 20

L
linear optimization, 135
linear program, 135

M
matrix, 41
median, 14
    definition, 15
    properties, 15
    weighted median, 84, 85
multicriteria evaluation, 38

© Springer International Publishing AG 2016
S. James, *An Introduction to Data Analysis using Aggregation Functions in R*,
DOI 10.1007/978-3-319-46762-7

N
negation, 42
    strict, 42
normalization, *see* transformations

O
ordered weighted average (OWA), 101
    definition, 101
    fitting, 144
    properties, 102
    special cases, 102
ordered weighted averaging (OWA)
    quantifier, 107
orness, 103
overfitting, 142

P
piecewise-linear transformation, *see*
        transformations
power mean, 57
    definition, 58
    fitting, 143
    special cases, 58
    weighted power mean, 82

Q
quantifier, 107
quasi-arithmetic mean, 60
    definition, 60
    fitting, 143

R
R
    AggWAfit
        f.plot3d, 21
        fit.choquet, 150
        fit.OWA, 150
        fit.QAM, 149
    arithmetic mean, 32
    Borda count, 94
    Choquet integral, 122
    defaults, 92
    expressions
        if, 68
    functions, 8
        abs, 132
        array, 29
        as.numeric, 66
        c, 29
        cbind, 64

        cor, 141
        cumsum, 93
        factorial, 123
        intToBits, 124
        length, 31
        lp, 135
        match, 94
        matrix, 65
        max, 31
        mean, 30
        min, 31
        order, 67
        prod, 31
        rank, 67
        rbind, 64
        read.table, 65
        rep, 29
        sd, 67
        seq, 29
        sort, 67
        sum, 31
        which.max, 93
        write.table, 66
    geometric mean, 33
    harmonic mean, 33
    median, 31
    Minkowski distance, 132
    negations, 67
    ordered weighted average (OWA),
        121
    orness, 124
    power mean, 69
    Shapley value, 125
    statements
        for, 70
    transformations, 67
    trimmed mean, 121
    weighted arithmetic mean, 91
    weighted median, 93
    weighted power mean, 92
    Winsorized mean, 121
rank-scaling, *see* transformations
recommender systems, 130
regression
    parameters, 78
reliability, 157

S
scaling, *see* transformations
Shapley value, 116
    definition, 116
Simpson's dominance index, 85
standardization, *see* transformations

T
transformations, 56
    linear feature scaling, 47
    log, 50
    normalization, 46
    piecewise-linear, 52, 53
    polynomial, 50
    rank-scaling, 49
    standardization, 46, 48
trimmed mean, 99

W
weighted averaging, 75
weighting vector, 79,
        104
    definition, 79
    interpretation, 81
    learning weights from data,
        143
    quantifier, 107
Winsorized mean, 99

Printed in the United States
By Bookmasters